An Introduction to Causes of Animal Diseases

An Introduction to Causes of Animal Diseases

Balaji Yadav Maddina

RANDOM
RANDOM PUBLICATIONS
NEW DELHI (INDIA)

An Introduction to Causes of Animal Diseases

ISBN 978-93-5111-882-4

Published in 2016 in India by

RANDOM PUBLICATIONS

4376-A/4B, Gali Murari Lal, Ansari Road
New Delhi-110 002
Phone : +9111-43580356, 011-23289044, 011-43142548
e-mail: sales@randompublications.com,
info@randompublications.com, randomexports@gmail.com

Reprinted 2021

Type Setting by : Friends Media, Delhi-110089
Digitally Printed at : Replika Press Pvt. Ltd.

Preface

Animal Diseases has been carefully compiled and edited to meet the long felt needs of increasingly large number of those who have to deal with the different aspects of human diseases in colleges, universities and research institutes. It provides a stimulating and important new view of interaction between animals and pathogens causing diseases.

Animal diseases that people can catch are called zoonoses. Many diseases affecting humans can be traced to animals or animal products. You can get a disease directly from an animal, or indirectly, through the environment. Farm animals can carry diseases. If you touch them or things they have touched, like fencing or buckets, wash your hands thoroughly. Adults should make sure children who visit farms or petting zoos wash up as well. Though they may be cute and cuddly, wild animals may carry germs, viruses, and parasites. Deer and deer mice carry ticks that cause Lyme disease. Some wild animals may carry rabies. Enjoy wildlife from a distance.

The control of animal diseases and the promotion and protection of animal health are essential components of any effective animal breeding and production programme. Despite remarkable technical advances in the diagnosis, prevention and control of animal diseases, the condition of animal health throughout the developing world remains generally poor, causing substantial economic losses and hindering any improvement in livestock productivity.

The book is intended to acquaint students of various fields involved directly or indirectly with the major principles of human diseases. The book may be helpful as well to practitioners and those engaged in medical research.

– ***Author***

Contents

1

Introduction

The control of animal diseases and the promotion and protection of animal health are essential components of any effective animal breeding and production programme. Despite remarkable technical advances in the diagnosis, prevention and control of animal diseases, the condition of animal health throughout the developing world remains generally poor, causing substantial economic losses and hindering any improvement in livestock productivity.

In developing countries, animal health services were established with the main objective of controlling major contagious and infectious diseases, such as foot-and-mouth disease, rinderpest and contagious pleuropneumonia, as well as parasitic diseases, such as trypanosomiasis and tick-borne diseases. This was obviously the first priority, since the control of these diseases is a prerequisite to any successful livestock development programme.

With the present concern for sustainable economic development, more attention is now being given to other diseases that affect livestock productivity, such as helminthiasis, nutritional diseases, reproductive disorders, etc.

The successful control of disease depends initially on its timely and accurate recognition and on the presence of sound diagnostic capabilities based on effective working links between laboratories and field services. Emergencies created by outbreaks of major infectious diseases demonstrate the need for establishing, strengthening and improving such diagnostic services. As well, particular attention should be given to the development of an efficient animal disease information system.

Beyond the national level, a general increase in the movement of animals and animal products underlines the importance of international cooperation in the prevention and control of animal diseases.

Most animal health services in developing countries do not have at present adequate technical and administrative infrastructures to carry out the tasks and duties necessary for the efficient control of animal diseases and for consumer protection.

In many developing countries, there is either a shortage of skilled veterinary personnel or their services are not correctly utilised. The problem is exacerbated by deficient veterinary infrastructures and inadequate disease control programmes, veterinary legislation and information services, as well as a lack of transport, communications, veterinary products and equipment. The most common problem is the shortage of funds to sustain the activities of veterinary staff. In some developing countries, the animal health services are not given the appropriate legal power in the administrative system.

These shortages significantly reduce the effectiveness of animal health services control measures against major animal diseases. By exploiting existing conditions, local resources and international help, possibilities for improving of animal health service programmes are increased.

Veterinary education and training should receive the highest priority; however, more emphasis must be placed on the qualitative and practical aspects. Effective personnel development requires adequate planning of staff requirements, improved curriculums for undergraduate and postgraduate studies, training of auxiliary personnel and greater interregional cooperation to exploit existing training facilities fully, among other things.

Animal health services should be further developed by improving their efficiency. While the control and prevention of major infectious diseases clearly remains a government responsibility, some veterinary tasks such as the treatment of individual animals could be undertaken in other ways. Privatisation is one way of improving some sectors of animal health and of responding suitably to the needs of animal owners.

Other ways include contracting out certain services, creating farmer cooperatives and producer associations, recovering government costs more efficiently and using the revenue thus generated for selective subsidies. Several countries have already taken steps to reorganise their veterinary services in this way.

Developing countries must address this problem. This publication aims to assist them in improving livestock production through better control of major animal diseases. It is not intended to be a comprehensive text describing in-depth all aspects of a complex subject with worldwide variations. Instead, it is meant to serve as a guideline, providing general background information on basic topics. Its main objective is to assist animal health authorities in their organisation, planning and management activities.

The contents cover major problems facing official animal health services in developing countries in contributing to the production of food of animal origin and livestock development as integral components of general social, economic and agricultural development. Other priorities of animal health services include the protection of humans against diseases that may be transmitted by animals and the production of safe food.

Biological and pharmaceutical production may be the responsibility of some services, however, only aspects of the control and management of veterinary biologicals and drugs are included in this publication. The major issues dealt with in this publication are the objectives, functions, organisation and management of animal health services. Without going into detail, relevant information, statements and recommendations make up the basic text. This general approach does not permit specific disease control or specific social, economic and ecological conditions to be dealt with. However, information is provided in the annexes on specific topics and in the selected bibliographies of general interest publications.

AVIAN INFLUENZA IN CATS

Cats with Avian Influenza exhibit symptoms that can result in death. They are one of the few species that can get Avian Influenza. The specific virus that they get is H5N1, which is a subtype of Avian Influenza. In order to get the virus, cats need to be in contact with waterfowl, poultry, or uncooked poultry that are infected. Two of the main organs that the virus affects are the lungs and liver.

HISTORY

The H5N1 virus has been found in China, Thailand, Vietnam, Indonesia, Iraq, Austria, and Germany. Besides being found in domestic cats, the virus has infected a variety of wild cats such as the Asiatic golden cat, the clouded leopard, tigers, and leopards.

H5N1 was first discovered in domestic and wild cats in Asia, specifically in 2003 in the Thai zoo where two tigers and two leopards died. In 2004, the Thai zoo had 147 tigers that died or were euthanized. This was then followed by an outbreak in Germany in 2006, where three stray cats were found to be either dying or dead during the peak time of the virus outbreak.

Because the virus infects the lungs of cats, it is one of the preferred model animals to study the effects of H5N1 in humans.

VIRUS

The most common way a cat can obtain H5N1 is by consuming an infected bird. This has been studied in the 2006 and 2007 cases in Germany and Austria where the strains between the cat and the infected birds were not different between the species. A cat is able to then transfer the virus via the respiratory tract and the digestive tract to other cats. However, studies suggest that a cat cannot transfer the virus to a dog, and vice versa, while sharing a food bowl. Though there is no concrete evidence, there is a potential link between the transfer of the virus between poultry, wild birds, and humans.

Once the cat is infected, after an incubation period of 2 to 3 days, the virus can be found in the respiratory tissues, attached to the type II pneumocytes and alveolarmacrophages, as well as the intestinal tissues. There have also been some cases where the virus has been found in the brain and other systems in the body.

As of right now the H5N1 virus has not adapted to transfer in between mammals, but there is a fear that this can occur.

Studies done with cats

One epidemiological study that was performed in Germany and Austria on 171 cats found that less than 1.8% of this population had H5N1. In this same sample population of cats, less than 2.6% had antibodies to H5N1. Even though Germany and Austria are among the countries that have had naturally occurring cases, this study shows that very few cats have contracted the disease.

There have also been studies looking at the T cells, specifically CD4 and CD8, in the cat after viral infection. Though the mechanism is not fully known, there seems to be an inverse relationship with the amount of T cells present and the amount of infected cells.

Another study to test whether the ALVAC recombinant canarypox virus could prime the immune system in cats was performed. This vaccine has the same hemagglutinin as the H5N1 virus, and therefore worked on preventing death from two different strains of the virus, HPAIV A/Vietnam/1194/2004 and HPAIV A/Indonesia/05/2005. However, some of the cats that were vaccinated did exhibit hyperthermia and weitht loss, and all of the cats did have some disease change (assuming lesions) in their lungs. All of the cats, except one, still excreted the virus even after being vaccinated.

SYMPTOMS

A cat that is infected with a high dose of the virus can show signs of fever, lethargy, and dyspnea. There have even been recorded cases where a cat has neurological symptoms such as circling or ataxia. In a case in February 2004, a 2-year-old male cat was panting and convulsing on top of having a fever two days prior to death. This cat also had lesions that were identified as renal congestion, pulmonary congestion, edema, and pneumonia. Upon inspection, the cat also had cerebral congestion, conjunctivitis, and hemorrhaging in the serosae of the intestines. However, a cat that is infected with a low dose of the virus may not necessarily show symptoms. Though they may be asymptomatic, they can still transfer small amounts of the virus.

TREATMENT AND PREVENTION

Cats can be protected from H5N1 if they are given a vaccination, as mentioned above. However, it was also found that cats can still shed some of the virus but in low numbers. If a cat is exhibiting symptoms, they should be put into isolation and kept indoors. Then they should be taken to a vet to get tested for the presence of H5N1. If there is a possibility that the cat has Avian

Influenza, then there should be extra care when handling the cat. Some of the precautions include avoiding all direct contact with the cat by wearing gloves, masks, and goggles. Whatever surfaces the cat comes in contact with should be disinfected with standard household cleaners. They have given tigers an antiviral treatment of Oseltamivir with a dose of 75 mg/60 kg two times a day. The specific dosage was extrapolated from human data, but there hasn't been any data to suggest protection. As with many antiviral treatments, the dosage depends on the species.

CANCER IN DOGS

Similarly to humans, cancer is the leading cause of death among older dogs. It accounts for approximately 50% of deaths each year but can be successfully treated if diagnosed early. The medical science that studies cancer in animals is called veterinary oncology and veterinarians that specialize in cancer diagnosis and treatment are called veterinary oncologists. As with people, dogs can get various kinds of cancer. Cancer in dogs is one of the leading cause of deaths in dogs but it can be successfully treated.

CAUSES

Cancer is a multifactorial disease, which means it has no known single cause. However, it is known that hereditary and environmental factors can contribute to the development of cancer in dogs.

SYMPTOMS

Symptoms of cancer in dogs may include:

- Lumps (which are not always malignant, but should always be examined by a vet)
- Swelling
- Persistent sores
- Abnormal discharge from any part of the body
- Bad breath
- Listlessness/lethargy
- Rapid, often unexplained weight loss
- Sudden lameness
- Offensive odor
- Black, tarry stools (a symptom of ulcers, which can be caused by mast cell tumors)
- Decreased or loss of appetite
- Difficulty breathing, urinating or defecating

SUSCEPTIBILITY

Older dogs are much more likely to develop cancer than younger ones, and certain breeds are prone to specific kinds of cancers. Boxers, Boston

Terriers and Golden Retrieversare among the breeds that most commonly develop mast cell tumors. Large and giant breeds, like Great Danes and Saint Bernards, are much more likely to suffer from bone cancer than smaller breeds. It is important for the owner to be familiar with the diseases to which their specific breed of dog might have a breed predisposition.

TREATMENT

Treatment options vary and depend on the type and stage of cancer. Common treatments include surgery, chemotherapy, radiation therapy and immunotherapy. A combination of therapies may be used. Success of treatment depends on the form and extent of the cancer and the aggressiveness of the therapy. Early detection offers the best chance for successful treatment.

Some dog owners opt for no treatment of the cancer at all, in which case palliative care, including pain relief, may be offered. Regardless of how treatment proceeds following a diagnosis, the quality of life of the pet is an important consideration. Some canine cancers can be cured, and almost all patients receive at least some benefit from treatment. In cases where the cancer is not curable, there are still many things which can be done to alleviate the dog's pain. Good nutrition and care from the dog's owner can greatly enhance quality of life.

COMPLEX VERTEBRAL MALFORMATION

Complex vertebral malformation or CVM is a lethal hereditary syndrome found in Holstein cattle. CVM is responsible for malformed calves that are either spontaneously aborted or die shortly after birth. It is caused by a missense mutation in the SLC35A3 gene. Since the mutant form of the gene is recessive, only individuals carrying two copies of the faulty gene (homozygous individuals) are affected. Heterozygous individuals, those who carry one copy of the faulty gene and one copy of the normal gene, have no symptoms but may still pass the disease on to their offspring.

ORIGIN

CVM has been traced back to the bull Carlin-M Ivanhoe Bell who lived in the 1980s. He was used for two decades in international Holstein breeding for the exceptional milk production he passed on to his daughters. Unusually, Carlin-M Ivanhoe Bell was a carrier for two genetic diseases, CVM and Bovine leukocyte adhesion deficiency (BLAD). The BLAD and CVM genes are located in different chromosomes. When the sire (father) of Carlin-M Ivanhoe Bell, a bull named Pennstate Ivanhoe Star, was tested he was found to be a carrier of both CVM and BLAD. Carlin-M Ivanhoe Bell's grandsire Osborndale Ivanhoe, however, carried only BLAD. Scientists therefore believe that the mutation responsible for CVM occurred either in Pennstate Ivanhoe Starissa, or somewhere in his maternal family.

The CVM syndrome was first found in a Danish Holstein stock in 1999, and during the following years it was also found in the United States, United Kingdom, Netherlands, and Japan. Because of the wide international usage of Carlin-M Ivanhoe Bell and the large number of animals descending from him, the CVM gene is found in Holstein cattle throughout the world. By the turn of the 21st century, more than 30 percent of the best Holstein sires in both Denmark and Japan were CVM carriers. A test for CVM was developed in the turn of the 21st century, and was first used in Denmark in 2000.

EFFECTS

CVM is caused by a missense mutation in the bovine SLC35A3 gene. The mutant protein has the amino acid phenylalanine at position 180 instead of valine. This causes abnormal nucleotide-sugar transport into the Golgi apparatus, leading to malformations of the vertebral column. CVM is the first genetic disorder found to be caused by defects in the SLC35A3 gene. The mutation responsible for CVM is recessively inherited, and heterozygous carriers of the mutation are asymptomatic. CVM analysis found the bovine SLC35A3 to be the first nucleotide-sugar transportation regulating gene also responsible in the formation of vertebrae and ribs.

CVM affects foetal development, being a cause of frequent abortions and stillbirths. Affected calves express low birth weight and a variety of malformations in the vertebrae and heart. Diagnose based on just visual examination of a calf may be difficult due to the wide variety in the expression of anomalies, and a definite diagnosis requires DNA testing.

Abortions

As many as 88 percent of homozygous, CVM-affected foetuses are spontaneously aborted within 260 days from the insemination. A normal bovine pregnancy lasts 280 days. Only 4-5 percent of CVM affected foetuses are calved alive. These calves however are nonviable.

Malformations

A CVM affected calf typically has an abnormally short neck and crooked pasterns. Other malformations associated with CVM are malformations or fusing of vertebrae,scoliosis, contraction and crookedness of distal joints, and abnormal shape of head. Hearts of CVM affected calves have been found to have, among other, abnormal placement of main vessels.

Production and breeding

CVM is a cause of multiple problems for a cattle keeper. A CVM carrier cow's fertility is 25% lower than normal, and abortions lead to lowered milk production. For these reasons, CVM carriers are often removed. CVM carrier bulls have been pre-eliminated in Europe since the turn of the 21st century. Older bulls were also tested, and some of the best animals were found to be carriers. The removal of this material led to a period of slowing in dairy breeding in a number of European countries.

PREVENTION AND CONTROL OF ANIMAL DISEASES

People who own or keep animals are responsible for their care and health. The Government has set a number of rules to monitor the health of kept animals. These rules largely concern the prevention, monitoring and control of animal diseases. The Government has also set specific rules for a number of animal diseases, for example because of their potential social and economic impact or because they pose a threat to human health.

ANIMAL HEALTH MONITORING

The Government constantly monitors animal health in the Netherlands. This facilitates rapid changes in animal health policy, or immediate response to outbreaks of infectious diseases.

The Government has set three objectives in consultation with the sector:

- To trace outbreaks of known diseases and pathogens that are not endemic (i.e. do not arise in nature) in the Netherlands;
- To identify as yet unknown diseases;
- To monitor trends and developments.

Prevention of animal diseases

To prevent the introduction and dissemination of animal diseases as far as possible, owners of animals must (among other things):

- Do their best to keep animal diseases out of their business;
- Ensure adequate hygiene;
- Be alert to symptoms of disease;
- Comply with requirements when importing animals from other countries;
- Notify a vet of any suspected animal disease.

When there is an increased risk of infectious animal diseases entering the country, more stringent supervision of animal transports is applied. These measures can be at farm or national level (e.g. a ban on imports from infected countries).

Travellers

There are some holiday destinations where infectious animal diseases, such as foot and mouth or bird flu, are prevalent. The Government therefore advises travellers to be careful about contact with animals abroad. Pathogens from these animals can be accidentally introduced into the Netherlands with travellers and their luggage. There are also rules for travellers in respect of the importation of animals and animal products such as wool, processed meat or dairy products or hunting trophies. Bringing live animals or animal products into the Netherlands is prohibited.

Notifiable animal diseases

Stricter rules apply to certain animal diseases in terms of reporting, identification and control. These diseases are on the list of Notifiable animal diseases (in Dutch). When such diseases are notified the Ministry of EL&I can announce additional control measures.

- The reasons for requiring notification of such animal diseases are as follows:
- They can spread quickly;
- They can cause serious damage to the species in question;
- They cannot be prevented or controlled by normal commercial methods;
- They present a serious threat to public health.

Animal diseases can also be designated as notifiable under international conventions.

Reporting animal diseases

If an animal shows symptoms of a notifiable animal disease, the owner, vet or laboratory must notify the authorities by telephoning the national desk for animal diseases: (045) 546 31 88. This 24-hour service is run by the Food and Consumer Product Safety Authority (NVWA). Prompt reporting of an infectious animal disease or its symptoms is crucial. Without early reporting a disease can spread rapidly with potentially massive consequences for humans and animals.

Highly infectious animal diseases

The State is required to combat certain notifiable animal diseases. That means that the Government must always take measures to counter the spread of such diseases. The Minister for Agriculture and Foreign Trade is responsible for this disease control, while the NVWA is responsible for implementation. These animal diseases are treated in accordance with European directives, which are elaborated in the national policy scenarios of the Ministry of Economic Affairs. The approach is regulated in various scenarios.

Supplementary measures

The Minister for Agriculture and Foreign Trade can take measures over and above the policy scenarios. He is advised by the Chief Veterinary Officer and the Expert Groups.

EXPERTS

There are currently four Expert Groups:

- on ruminants;
- on poultry;
- on pigs;
- on horses.

The experts are independent and are appointed by the Minister. The groups consist of epidemiologists, virologists, vets and (for example) experts in the field of wild flora and fauna.

Vaccination

There are two types of vaccination:

Preventive vaccination

Practically all farm animals are vaccinated against diseases. Poultry, cattle, pigs, sheep and goats are all routinely vaccinated against pathogens such as Newcastle disease in poultry and Q fever in goats. This helps to keep the animals healthy and has no harmful effect on their products (meat, milk, etc.).

Vaccination for disease control

In the case of certain highly infectious animal diseases such as Classical Swine Fever (CSF) and Foot and Mouth Disease (FMD), emergency vaccination can be used a control measure. This counters the spread of the disease and prevents sickness in animals exposed to the infection. The Ministry of Economic Affairs decides in consultation with the EU if and when vaccination should be deployed. The relevant procedure is set out in the Ministry's policy scenarios.

The Rutte Government is campaigning within the EU for emergency vaccination as a control measure during outbreaks of highly infectious animal diseases.

2

Animals Classification

INTRODUCTION

Animals are classified into two main groups, Invertebrates and Vertebrates. Invertebretes do not have a backbone, while vertebrates Humans, birds and mammals) do. The animal kingdom is further divided into 20 phyla, of which the following are of agricultural importance:

PHYLUM PROTOZOA

This phylum consists of unicellular free-living animals (one cell). Most of them live in water, either seawater, fresh water or soil borne water. They are nearly all microscopic.

The various species of Protozoans move in a number of ways:

- Pseudopodia (False feet)
- Cilia (Small moving hairs)
- Tails

Reproduction of protozoans occurs by means of binary fission or mitosis. Amoeba is one of the most common protozoans and moves by means of pseudopodia. One of the most agriculturally important species of protozoans is Babesia. This protozoan causes Red-water fever, a disease that affects 100,000 cattle a year.

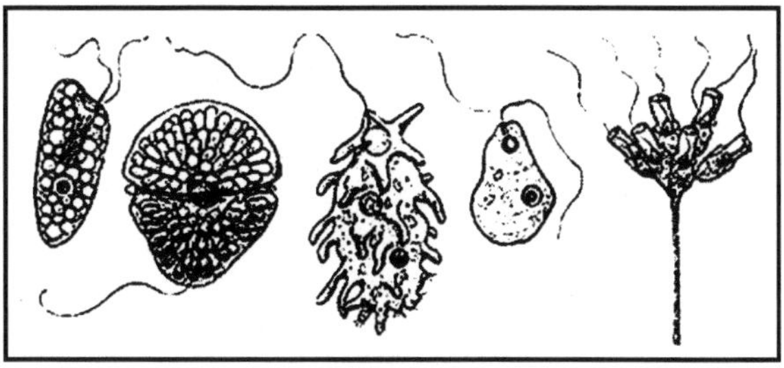

Ciliophora (ciliates, paramecia, spirotrichs):

- Cilia provide locomotion and feeding currents
- Rigid pellicle
- Two unequal nuclei

Sarcomastigophora (amebas, dinoflagellates, formaniferans, radiolarians):

- Unicellular or colonial protozoa
- Flagella and or pseudopodia provide locomotion
- Autotrophic or heterotrophic
- Homokaryotic (single kind of nucleus)

LABRYINTHOMORPHA (LABYRINTHMORPHANS)

- Complex network of ectoplasm extensions connecting the cells
- Apicomplexa (gregarines and coccidians, ex. Plasmodium responsible for malaria)

APICAL COMPLEX

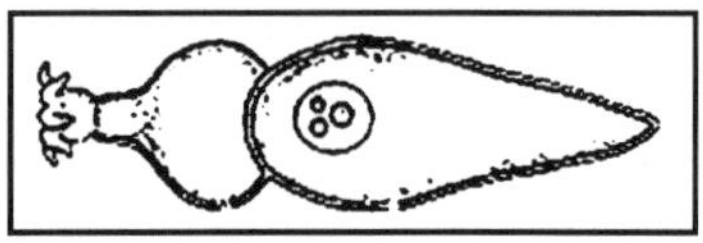

MICROSPORA (MICROSPORANS)

- Intracellular parasites
- Defined by spore structure (polar filaments)

ASCETOSPORA (ASCETOSPORANS)

- Intracellular parasites
- Defined by spore structure (no polar filaments)

MYXOZOA (MYXOZOANS)

Defined by spore structure

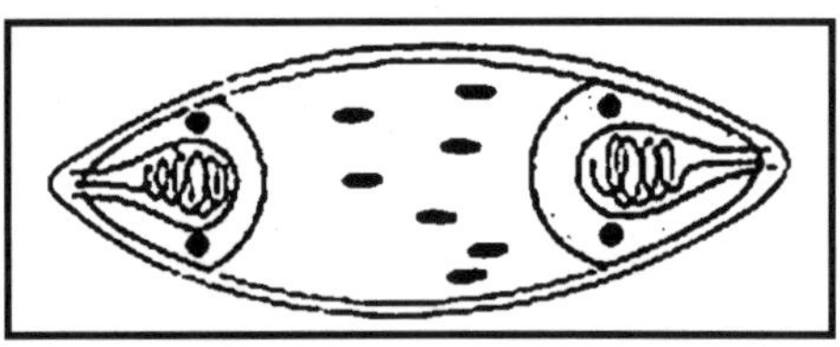

PHYLUM PLATYHELMINTHES

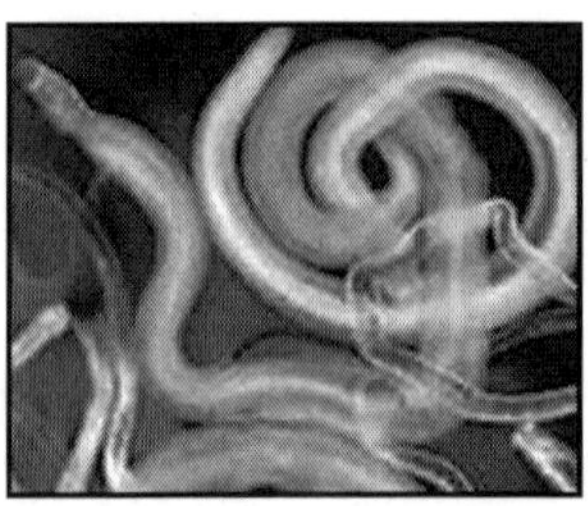

The phylum literally means "flat worms" and all the members of this phylum are flat worms!

The phylum is classified into three main groups or classes:

1. Class Turbellaria (Free living)
2. Class Trematoda (Parasitic Flukes)
3. Class Cestoda (Parasitic Tapeworms)

All platyhelminths are hermaphrodites and can, if required, can fertilize themselves. The most important agricultural species of platyhelminths is the Liver Fluke or Fasciola hepatica.

The life cycle of the liver fluke is very important in finding ways of controlling the disease. The life cycle is complicated and the fluke must lay huge amounts of eggs to survive. The life cycle takes place in the cow, on grass and in a secondary host (the mud snail).

THE LIFE CYCLE OF THE LIVER FLUKE

The Liver fluke lives in the ducts of the liver. The fluke lays eggs in the bile ducts (20,000 or so a day) The eggs pass in the faeces and hatch two weeks later in water and form a ciliated Miracidium. The Miracidium enters the foot of the mud snail and changes into a Sporocyst. Still inside the snail, the Sporocyst changes into a Redia. The Redia then produce very small tadpole shaped Cercaria. For every Miracidium that enters the snail, 10,000 Cercaria can be produced.

The Cercaria then leaves the snail and goes onto grass. There it becomes encysted (forms a shell) and waits to be eaten by a sheep or cow. If eaten, the stomach acids dissolve the cyst and the liver fluke moves to the liver and restarts the cycle.

Understanding the lifecycle of the liver fluke allows us control the spread in the following ways:

- Dosing any animals to kill the adult fluke
- Spraying molluscicides to kill the snail.
- Draining land (the snail only lives in water)
- Fencing flooded areas
- Don’t graze wet lands after August

PHYLUM NEMATODA

Nematodes are also known as roundworms or eelworms. There are huge numbers of species in this phylum.

All these worms reproduce by laying thousands of eggs, which become encysted in the grass and wait to be ingested.

The most important species are:

- Lungworms (Causes Hoose)
- Hairworms (Worms in school children)
- Potato eelworm
- Stomach worms

PHYLUM ANNELIDA

These are the segmented worms and include the earthworm – Lumbricus terrestris and leeches. On each segment, earthworms have four bristles or Chaetae, which they use to move.

Earthworms are important to the farmer because they improve the soil in the following ways:

- They eat their way through the soil and mix the ingested material with mucus in their guts. This helps to improve soil crumb structure.
- Depositing soil in different places and mixing horizons.
- Improve drainage of heavy clay soils
- Introduces more air into the soil.
- When they die they further increase the amount of organic matter.

PHYLUM MOLLUSCA

The molluscs include slugs, snails, squid, mussels, clams and octopus. These animals generally have a foot, which, excretes a slimy mucus.

They also have a rasping tongue. The most agriculturally important mollusc is the mud snail (Lymnaea truncatula)

PHYLUM ARTHROPODA

This is the largest phylum containing nearly a million species. Therefore it is necessary to sub-classify the phylum into classes. All Arthropods have jointed legs and an exoskeleton (outer skeleton). Members of the phylum include scorpions, insects, spiders, shellfish (crustaceans), woodlice, centipedes and millipedes. The most important classes of Arthropods are the spiders and insects.

CLASS INSECTA (INSECTS)

They have three main body parts: head, thorax and abdomen They include aphids, lice, fleas, crane-flies and butterflies.

The life cycle of all insects follows this path:

- Egg
- Larvae
- Pupa
- Adult

CLASS ARACHNIDA (SPIDERS)

The spider has two main body segments, the cephalothorax and the abdomen. Some of the diseases spiders cause on the farm are mange (scabies) and flea (mites).

Ticks are blood sucking spiders that can attack sheep and spread disease A lot of the members of phylum Arthropoda are parasites. Parasites are animals that live at the expense of other animals.

PHYLUM CHORDATA

All animals in the phylum Chordata have backbones and are vertebrates.

Examples include fish, birds, amphibians, reptiles and mammals. The two main agricultural classes are Class Aves (Birds) and Class Mammalia (Mammals).

Only these two classes are said to be homoeothermic – warm blooded – and can regulate their body temperature.

CLASS AVES

All members of this class have feathers, no teeth but a beak, lightened bones and no bladder.

CLASS MAMMALIA

All mammals have the following traits:

- Hair
- A Placenta
- Mammary glands to produce milk

Again they are further classified into sub-classes.

Some of the common sub classes are:

- Carnivores (Dogs and cats)
- Ungulates (Hooved animals)
- Odd toed - horse
- Even toed – sheep, cattle
- Marsupials – kangaroos
- Primates – monkeys and humans

PLACOZOA

- Marine
- Lacks anterior/posterior polarity and symmetry
- Cells arranged in double layer plate
- Flagella used for gliding motion
- Reproduces by fission

ORTHONECTIDA

- Asexual plasmodial phase and sexual ciliated phase
- In gonochoristic species, males attach to females

RHOMBOZOA

- Marine
- Outer layer of somatic cells surrounding core layer of reproductive cells
- Obligate symbionts in cephalapods

- Eutely (predetermined number of cells in each individual)
- Polar cap at the anterior end

MONOBLASTOZOA

- Body wall consists of single layer of cells bordering body cavity
- Bristles on "mouth" and "anus"
- Locomotion by ciliary gliding
- *Note*: There is only one record of this organism and it may not be accurate

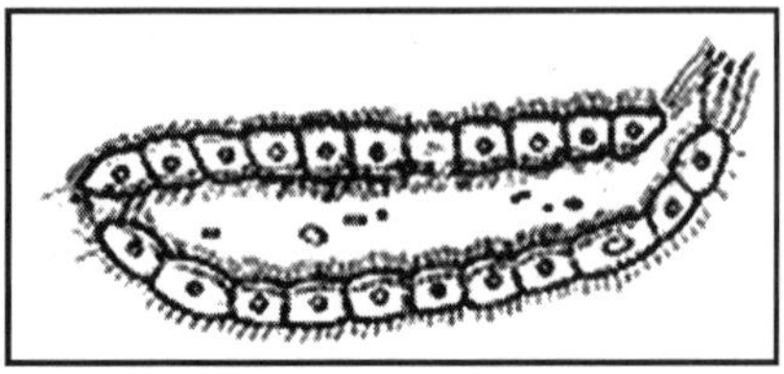

PORIFERA (SPONGES)

- Cellular level of organization (no true tissues)
- Cells totipotent (capable of developing into a complete organism or differentiating into any tissue in the organism)
- Choanocytes drive water through aquiferous system
- Sessile suspension feeders
- Some possess skeletal elements called spicules

CNIDARIA (JELLYFISH, SEA ANEMONES, CORALS AND HYDRA)

- Possess ectoderm and endoderm (diploblastic)
- Radial symmetry
- Possess Cnidae (commonly called nematocysts or stinging cells)
- Two-way gut
- No head, central nervous system, discrete gas exchange, excretory or circulatory systems
- Possess a nerve net

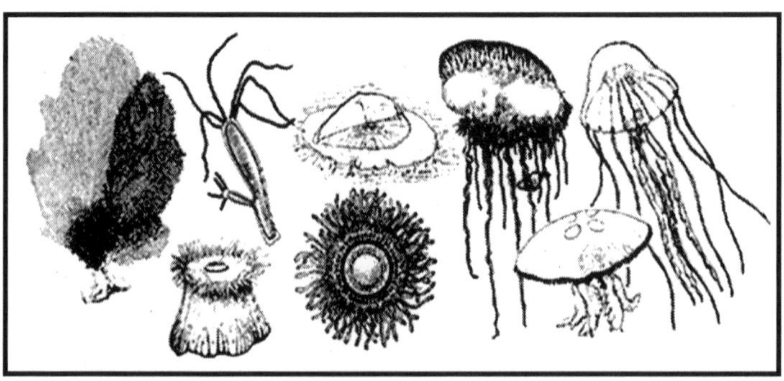

CTENOPHORA (COMB JELLIES)

- Diploblastic/li>
- Biradial symmetry
- Possess adhesive structures called colloblasts
- Gastrovascular cavity
- No head, central nervous system, discrete gas exchange, excretory or circulatory systems
- Possess a nerve net more specialized than cnidarians
- Eight rows of ciliary plates (combs/ctenes)
- Pair of long tentacles, usually retractable
- Typically hermaphroditic

PLATYHELMINTHES (FLATWORMS)

- Unsegmented worms
- Triploblastic, acoelomate
- Bilateral symmetry
- Complex but two-way gut
- Cephalized nervous system
- Possess primitive excretory and osmosregulatory structures
- Typically hermaphroditic

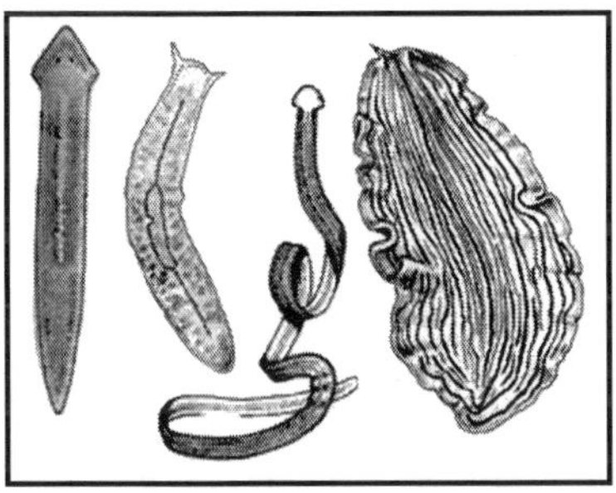

NEMERTEA (RIBBON WORMS)

- Unsegmented worms
- Triploblastic, acoelomate or coelomate
- Bilateral symmetry
- Complex one-way digestive tract
- Cephalized nervous system and closed circulatory system
- Possess unique proboscis in the rhnychocoel
- Typically gonochoristic

GNATHOSTOMULIDA

- Triploblastic, bilateral, vermiform, unsegmented and acoelomate
- Incomplete gut

- Unique jaw structure
- Hermaphroditic

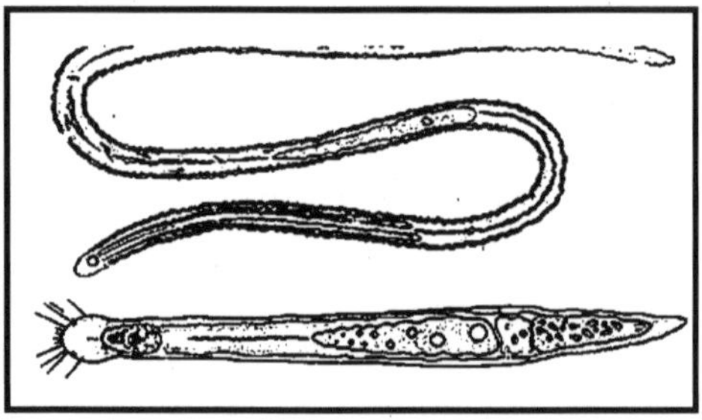

KINORHYNCHA (KINORHYNCHS)

- Triploblastic, bilateral, pseudocoelomate
- Body divided into 13 segments called zonites
- Molt
- Complete gut and protonephridia
- Gonochoristic

NEMATODA (ROUND WORMS)

- Triploblastic, bilateral, vermiform and pseudocoelomate
- Body covered in cuticle which must be molted
- Cephalic sense organs called amphids
- Six lips with sense organs surround mouth
- Unique excretory system but no special circulatory or gas exchange systems
- Gonochoristic

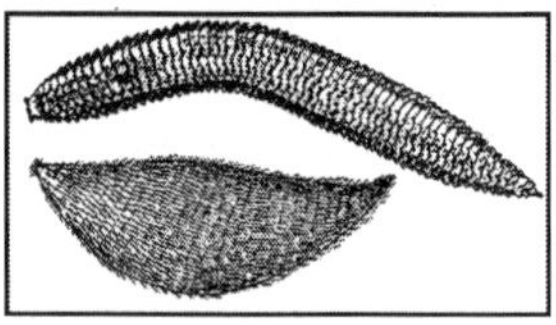

NEMATOPHORA (HORSEHAIR WORMS)

- Triploblastic, bilateral, vermiform and pseudocoelomate
- Body very long and thin
- Gut reduced or non-functional
- Larvae parasitic in arthropods
- Gonochoristic

ROTIFERA (ROTIFERS)

- Triploblastic, pseudocoelomate
- Bilateral symmetry

- Pharynx modified into mastix with jaw-like structures
- Ciliated corona for feeding
- Toes and adhesive glands at posterior
- Typically parthenogenetic (only females, asexually reproduce)

GASTROTRICHA (GASTROTRICHS)

- Triploblastic
- Bilateral symmetry
- Adhesive tubes
- Cuticle forming plates and spines
- Complete gut and protonephridia
- Hermaphroditic or parthenogenetic

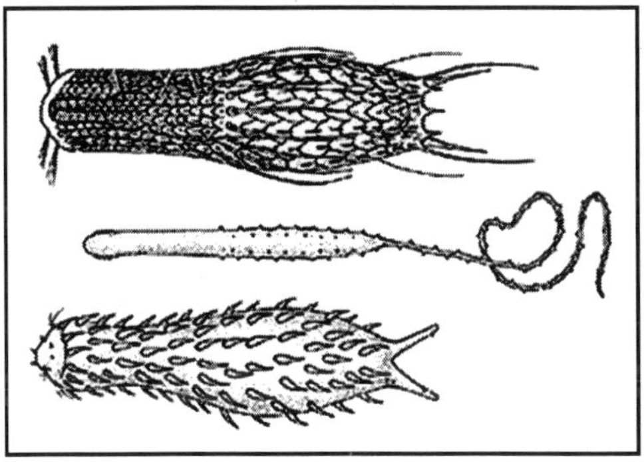

PRIAPULA

- Triploblastic, bilateral, unsegmented and vermiform
- Unique caudal appendage for gas exchange
- Complete gut
- Thin cuticle
- Gonochoristic

ACANTHOCEPHALA

- Triploblastic, bilateral, vermiform and pseudocoelomate
- Hooked proboscis at anterior end
- Channel system on epidermis called the lacunar system
- Gut absent
- Unique ligaments partition the body cavity
- Gonochoristic

ANNELIDA (SEGMENTED WORMS)

- Schizocoelomate, bilaterally symmetrical
- Segmented
- Complete digestive tract with regional specialization

- Closed circulatory system
- Cerebral ganglion and ventral nerve cord
- Paired, bundled setae on each segment

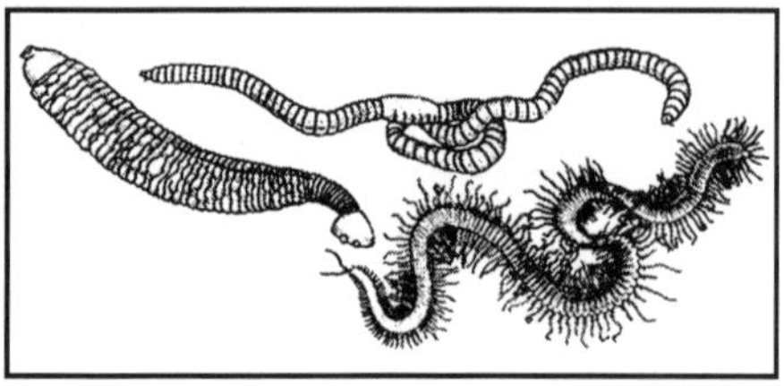

ECHUIRA

- Schizocoelomate, bilaterally symmetrical
- Unsegmented
- Complete digestive tract
- Simple closed circulatory system
- Non-retractable preoral proboscis

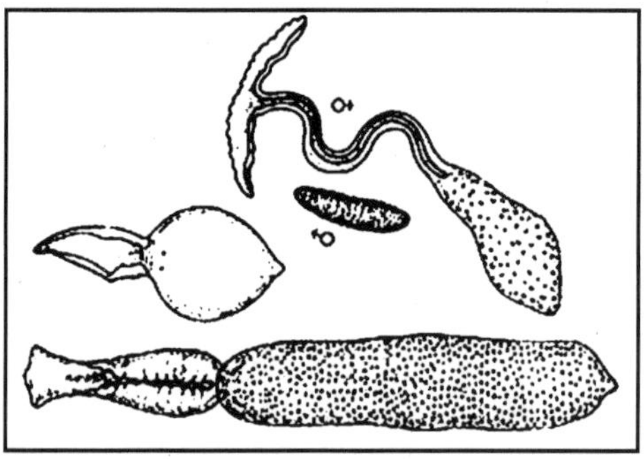

SIPUNCULA (PEANUT WORMS)

- Schizocoelomate, bilaterally symmetrical
- Unsegmented
- Complete U-shaped gut
- Mouth with tentacles
- Unique retractable anterior body region called introvert

ENTOPROCTA

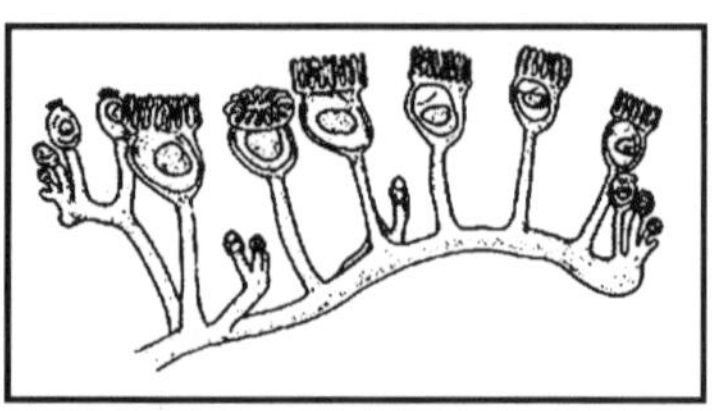

- Triploblastic, bilateral and unsegmented
- Sessile, stalked zooids

- Cup-shaped calyx
- Ring of tentacles enclosing mouth and anus
- Gonochoristic

LORICIFERA

- Triploblastic, bilateral, and pseudocoelomate
- Body divided into head, neck, and thorax, all of which are retractable into the abdomen
- Cuticular lorica
- Mouth housed in oral cone with stylets
- Gonochoristic

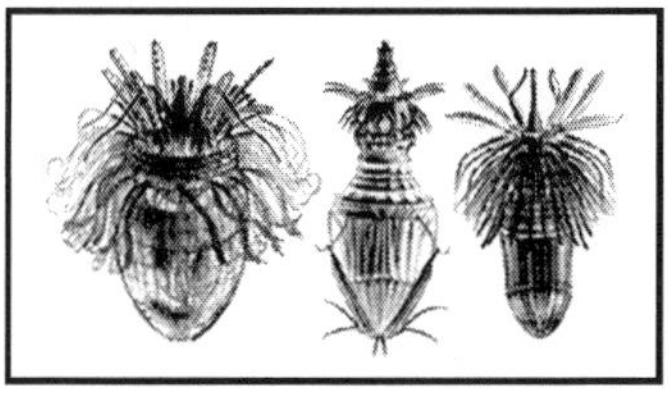

POGONOPHORA (BEARD WORMS)

- Coelomate, bilaterally symmetrical
- Reginally specialized body
- Lack digestive system
- Closed circulatory system
- Dwell in tubes
- Well developed anterior body cavities
- Three nerve cords

VESTIMENTIFERA (VENT WORMS)

- Coelomate, bilaterally symmetrical
- Reginally specialized body
- Lack digestive system in adults
- Closed circulatory system
- Dwell in tubes
- Vestimental wings
- Single nerve cord

MOLLUSCA (SNAILS, CLAMS, OCTOPUSES, TUSK SHELLS, CHITONS, ETC.)

- Coelomate, bilaterally symmetrical (secondary asymmetry)
- Principle body cavity a hemocoel
- Mantle covers body (thickened epidermis which secretes the shell, when present)
- Shell made of calcareuos material

- Muscular foot
- Buccal region with radula
- Complete gut

ARTHROPODA (INSECTS, CRUSTACEANS, SPIDERS, ETC.)

- Coelomate, bilaterally symmetrical
- Segmented body
- Hard exoskeleton (formed by cuticle, must be molted)
- Segmented, jointed appendages
- Pair of complex eyes and several simple eyes
- Open circulatory system
- Striated muscles

TARDIGRADA (WATER BEARS)

- Coelomate, bilaterally symmetrical
- Four pairs of clawed legs
- Ability to reduce metabolic activity during unfavourable conditions and form tuns (cysts) for their eggs
- Uncalcified cuticle, molting
- Smooth and striated muscle

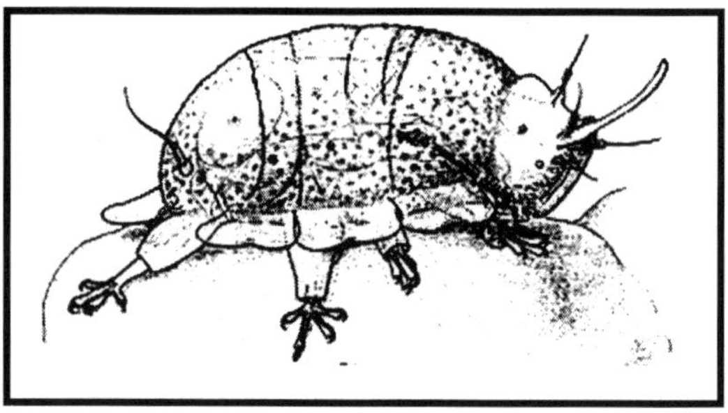

ONYCOPHORA (VELVET WORMS)

- Coelomate, bilaterally symmetrical
- Weakly hardened cuticle, molting
- Ciliated reproductive tract
- Reduced coelom and hemocoel
- Large brain

PENTASTOMIDA (TONGUE WORMS)

- Coelomate, bilaterally symmetrical
- Parasitic in lungs of vertebrates
- Leg like lobe appendages with claws
- Non-chitonous cuticle, molting
- Five snouts

BRACHIOPODA (LAMP SHELLS)

- Trimeric, coelomate
- Two shells enclose body
- Attach to substrate with a stalk called a pedicle
- Water filled mantle cavity
- Coiled lophophore
- U-shaped gut
- Open circulatory system

ECTOPROCTA/BRYOZOA (MOSS ANIMALS)

- Trimeric, coelomate, colonial
- Possess a a circular or U-shaped lophophore
- U-shaped gut
- Sessile colonies with polymorphic zooids (some specialized for feeding, some for reproduction, etc.)

PHORONIDA

- Worm like
- Build tubes out of sand grains
- Possess a lophophore
- U-shaped gut

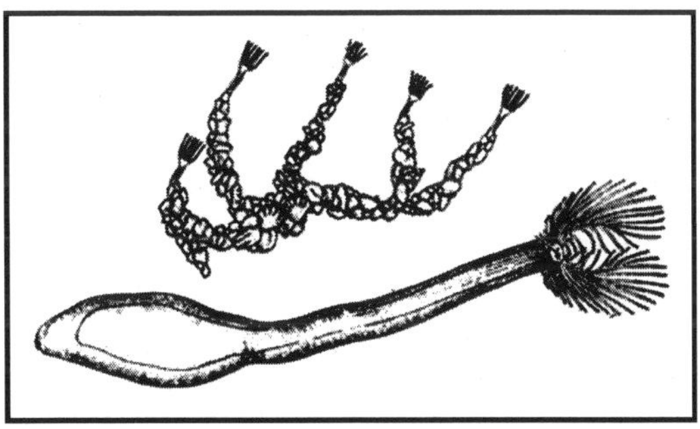

ECHINODERMATA (SEA STARS, BRITTLE STARS, URCHINS, SAND DOLLARS, SEA CUCUMBERS, CRINOIDS)

- Deuterostomic
- Calcareous exoskeleton made of embedded ossicles
- Pentaradial symmetry
- Water vascular sytem for locomotion
- Tube feet (podia moved by water vascular system)
- Complete gut (may be secondarily lost)
- No excretory organs
- Nerve net, nerve ring or radial nerves
- Ability to regenerate

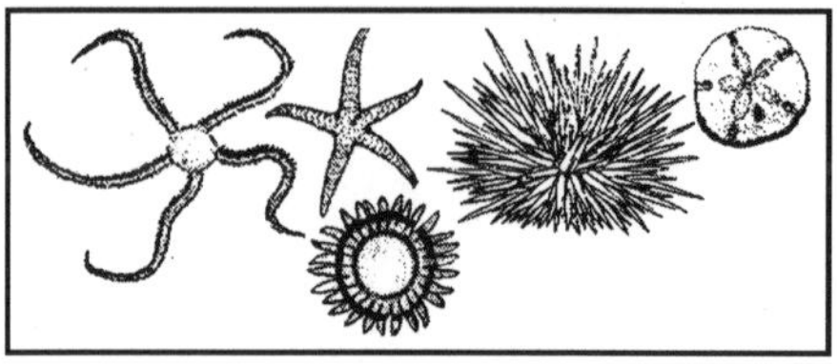

CHAETOGNATHA (ARROW WORMS)

- Bilateral deuterostome
- Long, steamlined body with head, trunk and post-anal tail
- Fins supported by rays
- Eyes, grasping spines and teeth on head
- No discrete excretory, circulatory or gas excganbge systems
- Large dorsal and ventral ganglia

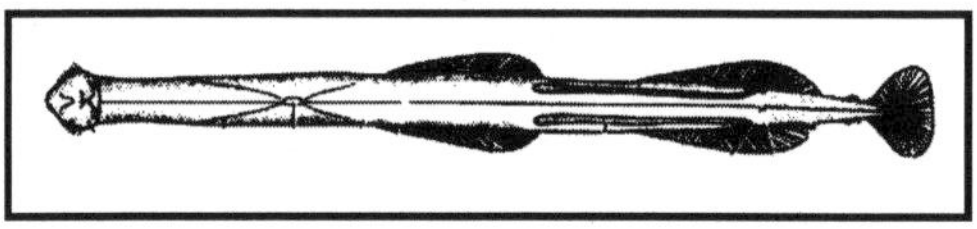

HEMICHORDATA (ACORN WORMS AND PTEROBRANCHS)

- Solitary, elongate, worm like
- Possess a stalked proboscis, collar and trunk
- Gill pores
- Tripartite regional division of the body

CHORDATA (TUNICATES AND VERTEBRATES)

- Bilateral deuterostome
- Notochord
- Hollow dorsal nerve tube
- Pharyngeal gills slits
- Postanal tail.

3

Animal Models of Disease

A common justification for the creation of GA animals is that they will provide, or contribute to, 'improved', more predictive models of disease. This is an oversimplification because a) this is not necessarily always true, and b) even where a new GA model is more appropriate, it may be used alongside other, older models by different researchers. There is no mechanism for ensuring only the most relevant are used and that newer models are available to all researchers. In such cases the GA animal is not the new *definitive* model but just an *additional* one. The motivation for the research may merely be an interest in the model for its own sake, but a medical application may be used to justify the work since this is likely to be more acceptable publicly and politically. Basic, fundamental research carried out within academic research establishments is not regulated under Directive 86/609, so it may not undergo an ethical review with appropriate assessment and weighing of harms and benefits. This directive is under review and is expected to be adopted end 2010. The revised directive may incorporate basic, fundamental research with proper review, however only if implemented and enforced properly.

There is also currently no requirement for GA animals to be cryopreserved and stored within central archive facilities or depositories. Such methods can reduce repetition and duplication of work by providing a central resource for use by the wider scientific community and protect against adverse events such as environmental disasters or genetic drift. They also reduce the need for live transportation of GA animals, with associated welfare problems, because frozen gametes or embryos could be sent instead.

GA ANIMALS IN TOXICITY TESTING

GA mice and rats are increasingly used in genotoxicity and carcinogenicity testing, within studies that are done to fulfil regulatory requirements for the marketing of chemicals and pharmaceuticals. Such animals are likely to suffer equivalent (or even more severe) levels of pain and distress to those experienced by animals in traditional tests, but it has been claimed that fewer animals will be needed.

Eurogroup believes that reducing the numbers of animals used in research and testing is an important goal, provided that this can be done without increasing the level of suffering experienced by individual animals.

However, relatively severe adverse effects have been reported in some GA strains used in toxicity testing. For example, mortality rates are higher in *c-neu* and *c-myc* mice used in carcinogenicity testing than in conventional mice used in the same type of test.

Many GA mice used in carcinogenicity tests are more susceptible to developing cancer and so will develop tumours more rapidly, which could make it more difficult to implement humane endpoints. Reducing numbers is thus not automatically a positive outcome for animals - the impact on individuals must be taken into account and it may be justifiable to use more animals who will suffer less.

In any case, the contribution that biotechnologies can make to reducing animal numbers in carcinogenicity and genotoxicity is not consistent, largely due to actual or potential regulatory requirements. In carcinogenicity testing, GA mice were introduced to try to reduce animal numbers and the time taken to obtain test results.

However, there are proposals to add additional control groups (a positive control and treated and untreated controls using non-GA mice) that would decrease the magnitude of the reduction if they are adopted. It is also uncertain whether testing on one GA strain will be regarded as sufficient by regulators, in which case the reductions will be further diminished by requirements for results obtained using other strains.

For genotoxicity testing, GA models have distinct scientific advantages over existing *in vivo* assays when used as second tier tests for *in vitro* genotoxins. The reduction in numbers of animals used would be fairly substantial, perhaps from 50 to 20 per substance tested, but again it is not certain that a test on one GA model would be regarded as sufficient. There are potentially much greater savings in animals if GA tests can be used in place of very large tests of heritable mutation which are currently used, albeit rarely, for chemicals of high concern.

Conversely, using GA animal models in toxicity testing could increase the use of animals by making some tests more practical, or by increasing the amount of information they produce.

For example, the use of GA animals could make investigations feasible that would otherwise require very large and impractical numbers of non-GA animals. Regulators might be inclined to ask for the GA test in cases where the conventional test would not have been requested (and could therefore presumably have been done without) because it was regarded as too cumbersome and possibly uninformative. Eurogroup believes that, in toxicity testing as in the other research fields discussed in this document, the focus should be on developing *in vitro* alternatives to replace animals and not on developing different animal models.

ANIMALS AS 'BIOREACTORS'

This category of GA animal use includes:

- "Pharmed" animals who produce therapeutic substances *e.g.* pigs who express the blood protein Factor IX in milk, and goats that express the anti-clotting agent Atryn in their milk;
- Animals producing specialist materials *e.g.* goats who produce spiders' silk proteins in their milk for use to make 'biosteel'. This has both medical and non-medical applications (*e.g.* in sutures and bullet-proof vests respectively) (Nexia Biotechnologies, Quebec);

The substance produced may have an adverse effect on the animal, either at the point of expression, or if it can enter the animal's bloodstream. For example, a strain of rabbits genetically engineered to express human erythropoietin (EPO) in the mammary glands also expresses the protein at low levels in other organs, resulting in greatly elevated numbers of red blood cells, infertility and premature death.

Using animals in this way, and referring to them as 'bioreactors', reinforces the perception of animals as units of production and/or biological tools, rather than as sentient beings with the ability to experience pain, suffering and distress.

XENOTRANSPLANTATION

The use of GA animals to supply organs, tissues or cells for transplantation into humans is a highly controversial issue which has been the subject of a great deal of debate. There are many legal, scientific, human health, animal welfare and ethical concerns, which have been described in a number of documents. Only the ethical and welfare issues relating to animals are addressed here.

These include:

- The ethics of genetically modifying animals of any species as a source of cells, tissues and organs for human transplantation;
- The harms associated with the initial creation of GA animals as source animals;
- The suffering and/or distress associated with production and maintenance systems for high health status source herds. This includes hysterotomy-derivation, early weaning practices, and barren husbandry environments, which have a serious negative impact on animal welfare because they prevent animals from satisfying their physical, social and behavioural needs.

Furthermore, development of xeno technology to a point where it can be used still requires a great deal of pre-clinical research. To date, such research has included studies of efficacy, physiology, immunology, and infection risks in a range of species including primates, goats and dogs. This research, by its very nature, causes considerable suffering. Experiments involving organ transplantation require major surgery, which in itself causes suffering that is exacerbated by tissue rejection and immunosuppressive treatment.

Xenotransplantation is also an example of a biotechnology where over-optimistic claims are made to justify the approach, the funding and the use of animals. For example, in September 1995, the UK company Imutran *"envisaged the first xenotransplants of transgenic pig hearts into human patients taking place in 1996"*. Yet despite some progress, particularly with cell transplants, the transplant of whole organs is no closer and xenografts still rarely survive for more than a few months.

CLONING COMPANION ANIMALS

Eurogroup believes that, without doubt, some applications of modern biotechnology are trivial, scientifically unnecessary and ethically unjustifiable.

Examples include;

- The cloning of champion racing and show jumping horses for sport;
- The generation of a green fluorescent rabbit 'GFP Bunny' as transgenic "art";
- The cloning of companion animals for example cats, purely to satisfy humans' emotional requirements.

CONCERNS RELATING TO AGRICULTURAL PRODUCTION

Gene mapping was/is the most widely used technique in agriculture to enhance selective breeding schemes, however as more and more 'super' animals exist this is being overtaken by the import/export of embryos and/or sperm from cloned 'super' animals. Examples of agricultural applications of gene mapping and other genetic altering techniques that have given rise to ethical and animal welfare concerns include:

INCREASING PRODUCTIVITY OR CHANGING BODY COMPOSITION

Gene mapping has been used primarily to select individuals for breeding with the aim of increasing productivity *i.e.* growth rates, litter sizes and production traits such as egg laying, milk volume and meat quantity and quality (including the proportion of lean meat to fat). Whilst still being practiced to create 'super' individuals for subsequent breeding, gene mapping is being/has been replaced by the import/export of embryos and/or sperm from cloned 'super' livestock (primarily from the US).

Farmed species have also been genetically manipulated to alter the composition of meat and milk. For example, pigs and cows respectively have been genetically altered to have higher levels of Omega 3 in their muscle, and to express higher levels of casein in their milk.

INCREASING DISEASE RESISTANCE

Animals are genetically manipulated to be resistant to disease, for example, cattle have been engineered to express the antibiotic lysostaphin in their milk, which results in increased resistance to mastitis. Animals (including pigs, sheep, mice and rabbits) have also been modified to express antibodies providing immunity to specific diseases, for example mice have been generated with protection against prion disease.

MAKING ANIMALS MORE 'ENVIRONMENTALLY FRIENDLY'

An example of this application is the 'Enviropig', which has been engineered to contain the enzyme phytase in the pigs' saliva so that they can digest sources of dietary phosphorus. This results in faeces with a lower phosphorus content, which in turn reduces the pollution of surface and ground water with phosphorus.

ETHICAL AND ANIMAL WELFARE CONCERNS

The selective breeding of farm animals has been conducted for thousands of years, and has given rise to a number of welfare concerns. However, Eurogroup believes that the use of cloning or other GA technologies to speed this process, or to introduce genes that could never be incorporated into the genomes of farm animals by any natural process, is a serious ethical and welfare issue. Directly altering an animal's genome is viewed by many as an unacceptable assault on the integrity of the animal that is incompatible with the concept of respecting farmed animals, and minimising the harms that are caused to them for human benefit. These views are important and should be respected as a legitimate part of the debate on biotechnology and farmed animal welfare.

Eurogroup also questions the necessity of further increasing production in farm animals. In many cases, productivity is already pushing animals to their physical and metabolic limits, so with any further increase there is an enhanced likelihood of animal welfare problems. Enhancing selective breeding, by gene mapping or genetic modification, can also cause suffering if the trait that is selected for has a negative impact on the rest of the animals' physiology. For example, hens who produce high numbers of eggs suffer from osteoporosis because the majority of the calcium they ingest is used in eggshell production.

There can also be less direct effects on welfare, in that some GA animals may receive lower standards of husbandry than conventional animals. For example, clinical mastitis is a major welfare problem in dairy systems with sub-optimal standards of hygiene, and early detection is reliant on routine inspections by parlour staff at milking. The creation of cattle resistant to mastitis may encourage the perception that mastitis is no longer a problem. This could not only compromise standards of parlour hygiene, but may reduce the level of attention paid to each animal at milking. This would increase the potential for other clinical or welfare problems to go undetected. "High productivity" animals may also be at risk if their husbandry is not appropriate. It may be possible for them to be properly managed and cared for in the controlled environment of a breeding company or experimental farm, but there are serious concerns regarding the welfare of such animals once they are released into commercial agriculture.

CONCLUDING REMARKS

Modern biotechnologies have had, and will continue to have, a serious adverse impact on animals, particularly with regard to their use in scientific research and agriculture. This adverse impact relates to the numbers of animals used and the nature of the harms caused to them. In addition, directly altering an animal's genome, as occurs in many applications of biotechnology, is viewed by many as altering the integrity of the animal in a way that is incompatible with the concept of respecting animals, and minimising the harms that are caused to them for human benefit. Lastly, the technology is progressing at a rate that is outstripping public understanding and ethical and public debate.

The development and application of novel biotechnologies therefore poses new challenges for existing regulatory regimes in a number of fields of science, medicine, agriculture and the environment. Eurogroup believes that the broader issues surrounding the ethical and social acceptability of such uses of animals, as set out in this submission, cannot be effectively addressed within the current regulatory systems. The following principles are fundamental to ensuring that the lives and welfare of animals involved in all modern and future biotechnologies are awarded due priority.

- It is critically important that all relevant regulatory systems are updated to take into account the animal welfare, ethical and social implications and societal concerns of the development and intended use of all modern biotechnologies.
- Biotechnology is applied in many different fields so there needs to be effective liaison, co-ordination and definition of responsibilities within and between all the relevant legislative and regulatory bodies concerned with a particular issue. This includes, for example, the different bodies regulating use of animals in experiments and those setting requirements for product regulation.
- The regulatory framework for each technology must encompass a process which enables a critical scrutiny of the potential harms to animals and the intended benefits, and a careful and fair weighing of these. This applies to broad research directions as well as individual projects and the further application of new technologies that result from these. Critical assessment of justification needs to be done prior to the development and/or application of a technology and must then

be reviewed regularly to check whether the harms and benefits are as expected so that appropriate action can be taken if necessary.

- A mechanism should be set in place to ensure that the justification and clinical relevance of all research involving the production and use of GA animals is critically scrutinised, such that animals are not used simply because the technology is available. This also needs to ensure that animal models of disease are regularly reviewed, so that redundant models are no longer routinely used for research purposes.
- There needs to be greater transparency with regard to the use of animals in biotechnology throughout Europe. Clearer information on the numbers of animals, and nature and level of any suffering that they experience, is essential in order to be able to identify issues of concern and assess trends. It is also vital that the public is well informed and therefore able to engage in constructive debate on the associated ethical issues.
- Restrictions should be placed on the species of animal that it is permissible to genetically modify. Non-human primates should not be genetically modified or cloned for any purpose, nor used as source animals for cells, tissues or organs.
- The production of GA livestock where the intention is to modify traits such as increased lean to fat ratio, growth rate, or litter size should not be allowed, as levels of productivity are already causing serious welfare problems.
- There should be far greater effort devoted to developing and validating alternatives to animal use in all fields. There should be greater commitment to, and endorsement of, the principles of the Three Rs of reduction, refinement and replacement in animal experiments.
  Examples especially relevant to modern biotechnology are:
  - The development of GA germ cells for *in vitro* testing;
  - The production of drugs, proteins, or material by bacteria rather than 'bioreactor' animals;
  - Generating cells, tissues and organs for treatment or transplantation using a patients own cells, eg human bladders.

GENETICALLY MODIFIED IN ANIMALS

Several animal species have already been genetically modified, and at least eleven have been cloned, though some scientists doubt the health of those clones that survived to birth. Some of these efforts are commercial, either for agribusiness or for sale directly to consumers as pets. Others are scientific experiments, usually defended as advancing, directly or indirectly, the cause of medicine for humans. The cloning or genetic modification of pets serves no justifiable purpose. These efforts serve only to play upon one set of emotions-our affection for our companion pets-in order to desensitize another set of emotions-our repugnance at the idea of treating them as artifacts that can be "designed" and manufactured.

The modification of livestock and the possibility of cloned meat entering the food chain are very controversial. So is the possible use of genetically modified animals to "grow" either pharmaceutical products or organs for transplant into humans.

CLONING

There have been published reports of the following species being cloned: carp, sheep, mice, cattle, goats, pigs, cats, rabbits, mules, horses, rats, and a deer. Some closely related species have also been cloned (a banteng, a wild cow, and a mouflon, a kind of sheep).

A gaur, a wild ox, was cloned but died within two days. Attempts have also been made, without success, to clone monkeys, dogs, pandas, chickens, and at least two extinct species: the Tasmanian tiger and the woolly mammoth. The mammoth experiment used an elephant surrogate and tissue found in permafrost.

Three major British institutions announced in July 2004 that they were setting up a tissue bank to preserve the DNA of endangered species, even after their extinction. Future cloning is seen as a possibility.

The Roslin Institute, where the first mammal was cloned, maintains records of all published mammalian cloning experiments up to July 2002 - 50 papers detailing 68 experiments, with 386 surviving clones. They conclude that the "overall efficiency of cloning is typically between 0 and 3 per cent (number of live offspring as a percentage of the number of nuclear transfer embryos), irrespective of the species, the donor cell type or technique." There is no evidence that efficiency has significantly improved since.

CLONING FOR LIVESTOCK

About 300 bulls have been cloned, with the aim of improving the quality of breeding stock. Cloning is too expensive, at around $20,000 per bull, to clone directly for meat, but breeding bulls are worth much more than that. The meat and milk of cloned cattle is presently kept off the market pending FDA approval. Nevertheless, BIO is actively promoting the use of cloned and transgenic animals for human food, as well as for pharmaceuticals.

The FDA is investigating the possibility of selling meat from cloned pigs as well as cattle, and in 2003 released a draft executive summary well in advance of the report it was supposed to summarize. "The FDA wants to gauge public reaction to the prospect of food from cloned farm animals before it decides whether to require government approval of the products before they are sold. That decision is expected to take another year."

PET CLONING

In 1998 Arizona billionaire John Sperling gave $3.7 million to Texas A&M University to clone his pet dog, Missy (the "Missyplicity Project"). Sperling is controversial. He became wealthy as founder of the University of Phoenix, a distance learning university that has been accused by many of being a "diploma mill." He has since used his assets to support a number of idiosyncratic endeavors.

The dog-cloning effort failed, and the team, reportedly against Sperling's wishes, began parallel efforts to clone a cat. The birth of the first cloned domestic cat was announced by the Texas A&M research team in February 2002. Born on December 22, 2001, "CopyCat" or "CC" was produced by fusing an ovarian tissue cell from an adult cat with a cat egg, and implanting the clonal embryo into an adult cat. This cloning "success" occurred after 188 failed attempts.

Sperling retained Lou Hawthorne, a publicist from Marin County, California, to handle public relations for his dog-cloning effort. Hawthorne subsequently established "Genetic Savings and Clone" as a profit-making venture to provide pet cloning services.

The company is offering to clone cats for $50,000. They also offer "gene banking" services for up to $1395, plus up to $150/year, all of which can be credited against the future cost of cloning. They announced in April 2004 that five customers had paid the fee, work had already begun on three other cats for staff members and one more slot was available, to make "nine lives." They hope to increase production to several thousand a year, and to reduce the cost substantially.

SCIENCE, MICE AND PATENTS

The most commonly genetically modified animal is almost certainly the mouse. It is small, short-lived and sufficiently similar to humans to be an almost ideal laboratory animal. As a result, mice have not only been cloned and modified, they have led to an actual industry in the production of "knockout mice," that is, mice with a particular gene or set of genes inactivated for research purposes. Trans Genic Inc asserts, "Currently, we are able to produce almost 1,000 strains of Knockout mouse in a year."

"Pharming". Cattle, sheep, goats, chickens, rabbits and pigs have been genetically modified with the aim of producing human proteins that are useful, generally as medicines. The gene transfer process is typically very inefficient, and cloning is seen as another way of propagating the GM animal.

A 1999 USDA report cited estimates that there was a $24 billion market for human proteins, and theoretically 600 transgenic cows could supply the worldwide demands for some drugs. In practice, however, several companies that have pursued this line have gone bust, and the profit potential seems less than it once did.

Genetic modification of animals in order to improve the prospects of organ transplants is also being investigated.

GENETICALLY MODIFIED FISH AS PETS

A tropical fish genetically modified to glow in the dark went on sale in Taiwan in 2003 for about $17 each. A different variety of zebrafish, called "GloFish," which were created in Singapore, reached the United States market in January 2004.

The distributor says that GloFish were originally developed to fluoresce only in the presence of pollutants, but that is not the form in which they are being sold. They cost about $5 each, and are intended to live in aquariums, but can breed and, in the right conditions, live in the wild.

The Food and Drug Administration (FDA) approved the sale without ceremony. A coalition led by the Center for Food Safety filed suit against the decision, but sales went ahead. In California, the Fish and Game Commission initially banned the fish but later agreed to hold hearings at the request of the distributor.

ALLERGY-FREE CATS

A company called Transgenic Pets, in Syracuse, NY, was widely reported in 2001 to be working with scientists at the University of Connecticut to "remove the allergen gene" from cats. The company hoped to raise $2 million and sell the modified animals for $1,000 each. Funding problems ended the project.

GENETIC MODIFICATION FOR AQUACULTURE

Most discussion of bioengineered food focuses on plants, but work on animals and fish is well under way. Genetically modified salmon have already been created, though FDA approval is not expected before at least 2006 and the target date has repeatedly been delayed. The Biotechnology Industry Organization (BIO) and the developers, Aqua Bounty Farms, claim that the GE salmon grow faster but not larger than ordinary salmon, but this is strongly disputed. Opponents also cite studies showing dramatic population crashes and unpredictable environmental impacts.

Salmon are not the only fish species being modified. Acting FDA Commissioner Lester M. Crawford summarized the position in a speech given in March 2004: "Less well known is that catfish and tilapia have been also genetically modified to grow faster and more efficiently than their non-transgenic counterparts.

Rainbow trout has been engineered to increase its contents of omega 3 fatty acids, and shellfish is being modified to reduce its allergenicity and make it grow faster."

The matter-of-fact, if not approving, tone of these comments is disturbing, as it comes from the head of the agency that is supposed to be regulating these technologies. Other countries, including Canada, have moved much faster than the United States to ban or at least place a moratorium on the genetic modification of fish. Malcolm Grant, Chair of the UK Agriculture and Environment Biotechnology Commission, points out that

"Once the fish has escaped, there's virtually nothing that can be done to recall it... Genetic biotechnology has opened a new chapter in the relationship between man and animals. We must therefore prepare now for developments that may be many years away." He also argues that pet cloning is "trivial, distasteful and could be sold to gullible owners."

GENETICALLY MODIFIED FARM ANIMALS

Acting Commissioner Crawford continued: "Cows can be bioengineered to produce several varieties of milk: milk with a lower level of a protein that's allergenic to some infants; milk that is more easily digested by people who are lactose intolerant; milk that has more naturally occurring antimicrobial enzyme, and therefore has longer shelf life; and milk that makes better cheese because it has altered distribution of caseins or less fat."

GENETIC MODIFICATION FOR LIVESTOCK

Research is also underway to use genetic modification to improve the health of cows and pigs. Advocates hope to produce cattle that would be resistant to Mad Cow disease, for example. There have also been reports of cows being genetically modified to "produce milk with higher than normal levels of protein, which would speed the process of making cheese."

CHIMERAS

Some researchers have created embryos with genetic material from both a human and an animal, also known as chimeras. Most claim this is for research purposes only, and such embryos would never be implanted or brought to term. But some bioethicists are unwilling to draw a line that would prevent it. For example, Jason Scott Robert and Francoise Baylis assert that they take "no stance at all" on whether "interspecies hybrids or chimeras from human materials should be forbidden or embraced." But much of their article is devoted to their contention that "the arguments against... creating novel part-human beings... are largely unsatisfactory."

The prospect of such human-animal chimeras being born raises a number of troubling questions. Does such an organism have human rights? What if it were 99.9 per cent human and 0.1 per cent chimpanzee? What of the reverse situation? The mixing of species can occur at three different levels. First, some DNA from one organism can be inserted into another organism's genome. Most often, however, these are described as transgenic animals rather than chimeras. For example, scientists have produced transgenic mice which contain some human genes.

Another type of chimera involves a blastocyst containing components from multiple species. James Grifo has used this technique for fertility research. After the US federal government informed him he could no longer continue his experiments, his team moved to China. There, they removed the nucleus from a fertilized rabbit egg and inserted a nucleus from a human somatic cell. The resulting embryos, which still contained rabbit mitochondrial DNA, were allowed to grow for 14 days before destruction.

Finally, scientists have placed cells from one species, typically human stem cells, into a multicellular embryo of another species. In 2004, researchers were surprised to discover that injecting human stem cells into a pig embryo resulted in a fetus whose cellular components were intermingled down to the genetic level. Many cells contained chromosomal DNA of both pig and human origin.

OPPOSITION

Several environmental and animal rights organizations have expressed opposition to pet cloning, including Friends of the Earth and the Humane Society of the United States, the largest animal welfare organization in the country. Jeremy Rifkin from the Foundation on Economic Trends and Stuart Newman of the New York Medical College and the Council for Responsible Genetics have pursued a novel path of opposition. They filed a patent application with the US Patent and Trademark Office for chimeric embryos and animals containing human cells. They did not intent to create a "humouse." Instead, they tried to force the PTO to take a position and, if the patent is granted, to prevent chimeras from being developed.

CLONING AND GENETICALLY ALTERED ANIMALS

At Eurogroup for Animals, we believe that humans have a duty to safeguard animals' welfare and health and we oppose all human activity that causes distress or suffering. In particular in the area of animal breeding, we are of the opinion that animals genetic constitution should not be changed if doing so causes suffering to themselves or to future generations.

Genetic selection must always seek to maximise the wellbeing of the resultant animal. Animal breeding should support a diverse gene pool in order that the genetic resources to maximise wellbeing are sustained for future generations. Techniques and technologies used in animal reproduction must not be used if they cause distress or suffering.

For these reasons Eurogroup is opposed to the cloning of animals and would urge the EU to introduce an immediate ban on the cloning of animals for food production, and on the sale of imported food products from cloned animals and their offspring. Eurogroup is concerned that delaying a decision will mean that food from cloned animals will enter the supply chain uncontrolled and without consideration of the opposition of EU citizens and consumers. Under Article 13 of the Lisbon Treaty, in the EU it is a legal obligation to pay full regard to the welfare requirements of animals when formulating and implementing the Union's agriculture, fisheries, transport, internal market, research and technological development and space policies.

Our concerns are based on the fact that some of the developments and applications of animal cloning and other genetic alteration techniques:

- Involve procedures that may cause animals pain, suffering or distress;
- Use a very large number of animals;
- Encourage a wider variety of applications leading to increased animal use;
- Increase the perception of animals as commodities for human use and/or gain, such as research tools or units of production;
- Are progressing at a rate that is outstripping public understanding and ethical and public debate.

CONCERNS RELATING TO HUMAN AND ANIMAL HEALTH

The main applications of cloned and other genetically altered (GA) animals (and their offspring) that are of concern at the present time are:

- The creation of genetically manipulated (GA) and cloned animals;
- The subsequent use of these GA animals (and their offspring):
  - As livestock on farms (used to standardize/maximise quantity and reduce the costs of processing/producing animal derived food products *e.g.* milk, meat, eggs);
  - As disease models;
  - In fundamental research *e.g.* to understand gene function;
  - In toxicity testing;
  - As bioreactors to produce biologically active compounds for experimental and/or medical purposes;
  - As sources of cells, tissues and organs for xenotransplantation;
  - In the creation of cloned pets, sports animals and 'living art'.

However, any new biotechnology, or new application of existing technologies, that is developed and/or tested in animal models, or that causes the animals pain, suffering, distress or lasting harm is of serious concern to Eurogroup. Examples include tissue engineering and the development of nanotechnology generally. Production of cloned and other GA animals

NUMBERS OF ANIMALS USED

Use of animals in experiments is a matter of serious public concern and pressure to reduce numbers has contributed to a downward trend in some countries. Creation and use of GA animals is reversing this trend and there has been an exponential rise in the number of GA animals used in scientific procedures each year in some EU member states. For example, in the UK, the number of GA animals has risen from around 50,000 in 1990 to 900,000 in 2004. In Germany, Ireland, Finland, Sweden and the Netherlands, there is also increasing creation and use of GA animals.

GA animals can be produced by a number of different techniques, all of which are inherently wasteful with respect to the number of embryos that are manipulated relative to the number of GA offspring subsequently produced. Current estimates suggest only a 3-5 per cent success rate when generating new GA animals.

The remaining offspring are surplus to requirements and are killed. A large number of animals are therefore used to provide sufficient eggs or embryos for genetic manipulation, and to act as recipients and foster mothers for the manipulated embryos.

These animals will usually be killed, either before harvesting the eggs or embryos, or, in the case of recipients, once their young are weaned. Thus GA technology is wasteful of animal life and the ethical implications of such wastage are important and need to be acknowledged.

POTENTIAL FOR PAIN, SUFFERING OR DISTRESS DURING CREATION OF GA OR CLONED ANIMALS

The procedures used to produce GA and/or nuclear transfer cloned animals involve hormonal and surgical interventions that can cause pain, suffering and distress.

These include; superovulation, vasectomy, semen and embryo collection and embryo transfer. Many of the manipulated embryos die during gestation. This may also cause suffering, although this depends on the stage of development at which death occurs and the species involved.

ADVERSE EFFECTS AS A CONSEQUENCE OF GENETIC MODIFICATION AND CLONING

Where GA animals are created as models of specific diseases, they can experience a range of adverse effects associated with the condition in question, which can have profound effects on their health and welfare. However, being genetically altered does not necessarily compromise the welfare of individual animals, and indeed some GA animals are indistinguishable from their non-GA siblings.

It also needs to be acknowledged that adverse effects may only become apparent when animals are subsequently maintained in a less well defined or controlled, environment than that of the laboratory or experimental farm. Nevertheless, in many cases, genetic alteration can have a deleterious effect on animal welfare and the harms caused depend on a number of factors.

Studies have shown that some *in vitro* culture procedures carried out during genetic manipulation or cloning protocols can lead to unpredictable complications in the animals subsequently produced. An example of this is Large Offspring Syndrome (LOS), which affects sheep, mice and cows following nuclear transfer.

As well as causing complications for the mother during birth, LOS encompasses a range of debilitating pathologies for the offspring including malformations in the liver, brain and urogenital tract, immune dysfunction, placental abnormalities, stillbirth, fetal overgrowth, respiratory failure and circulatory problems.

A further concern is that there is currently no mechanism to ensure consistency in the training of personnel in specific GA technologies and procedures, or to ensure that experimental and other refinements are disseminated throughout the biotechnology community. This can have a profound influence on the levels of pain, suffering and distress experienced by animals undergoing these procedures. It can also have a negative impact on the success rates achieved and levels of animal wastage.

CONCERNS REGARDING MUTAGENESIS PROGRAMMES

Mutagenesis (using chemicals such as ENU, or physical mutagens such as radiation, to increase the natural mutation rate of DNA) is the quickest method for producing large numbers of GA mice. This technique is being widely applied both in Europe and beyond, with mutagenesis programmes aiming to explore the function of every gene in the mouse genome.

This project presents a number of animal welfare concerns:

- The number of animals involved in mutagenesis programmes is vast; *e.g.* around 35-36,000 for a 3 year project, or around 50 animals per mutant line established;
- The process of mutagenising animals has significant animal welfare implications for the animals involved, for example males require 12-14 weeks to recover their fertility following treatment with a mutagen. Furthermore, only around 50 per cent of mutagenised males are able to go on to sire offspring with the remainder being culled;
- The mutations induced are by their very nature unpredictable and the scientific usefulness of mutagenised animals cannot therefore be predicted, nor can the effect of any mutation on the health and welfare of the animals. This makes the justification for producing GA mice in this way highly questionable.

CONCERNS REGARDING KNOCK-IN AND KNOCK-OUT TECHNOLOGY

A more targeted approach to the production of GA animals uses DNA constructs inserted into the animals genome to either completely remove a gene of interest (knock-out), or to replace a given gene with an altered version (knock-in). This is done either in the whole animal, or within specific tissues of an animal (conditional).

The animal welfare concerns include:

- At the current level of efficiency, the numbers of animals used is high - at least 200 animals will be used in the production of a single GA animal;
- The site of insertion into the host genome and the number of copies of the DNA construct inserted cannot always be controlled unless ES cell manipulation is used. This can be a problem because the incorporation of the construct DNA at an incorrect location can result in the random inactivation of other genes or alterations in the expression of surrounding genes, both of which can impact on the health and welfare of the animals produced;
- When a knockout animal is generated the level, or pattern of expression observed for other genes can be altered to compensate for the lost gene. Thus any effects observed in the GA animal may not only reflect the loss of the gene of interest, which means that investigation of the effects is not always straightforward. The use of knock-out as an alternative to knock-in or conditional technology, therefore needs careful consideration and justification.

SPECIAL CONCERNS FOR NON-HUMAN PRIMATES

This submission relates to all animals used in biotechnologies, but Eurogroup has particular concerns about the application of such technologies to non-human primates. Macaques have already been produced by nuclear transfer cloning in the USA and recent technical advances show that the production of GA primates is currently increasing. Eurogroup firmly believes that the genetic modification of non-human primates should not be allowed for any purpose, given the ethical issues raised by such developments, the large number of animals required to produce each GA animal, and the associated potential for harms.

CURRENT CONTEXT OF GENETICALLY ENGINEERED ANIMALS

Genetic engineering technology has numerous applications involving companion, wild, and farm animals, and animal models used in scientific research. The majority of genetically engineered animals are still in the research phase, rather than actually in use for their intended applications, or commercially available.

COMPANION ANIMALS

By inserting genes from sea anemone and jellyfish, zebrafish have been genetically engineered to express fluorescent proteins — hence the commonly termed "GloFish." GloFish began to be marketed in the United States in 2003 as ornamental pet fish; however, their sale sparked controversial ethical debates in California — the only US state to prohibit the sale of GloFish as pets. In addition to the insertion of foreign genes, gene knock-out techniques are also being used to create designer companion animals. For example, in the creation of hypoallergenic cats some companies use genetic engineering techniques to remove the gene that codes for the major cat allergen.

Companion species have also been derived by cloning. The first cloned cat, "CC," was created in 2002. At the time, the ability to clone mammals was a coveted prize, and after just a few years scientists created the first cloned dog, "Snuppy". With the exception of a couple of isolated cases, the genetically engineered pet industry is yet to move forward. However, it remains feasible that genetically engineered pets could become part of day-to-day life for practicing veterinarians, and there is evidence that clients have started to enquire about genetic engineering services, in particular the cloning of deceased pets.

WILD ANIMALS

The primary application of genetic engineering to wild species involves cloning. This technology could be applied to either extinct or endangered species; for example, there have been plans to clone the extinct thylacine and the woolly mammoth. Holt et al point out that, "As many conservationists are still suspicious of reproductive technologies, it is unlikely that cloning techniques would be easily accepted.

Individuals involved in field conservation often harbour suspicions that hi-tech approaches, backed by high profile publicity would divert funding away from their own efforts." However, cloning may prove to be an important tool to be used alongside other forms of assisted reproduction to help retain genetic diversity in small populations of endangered species.

FARM ANIMALS

As reviewed by Laible, there is "an assorted range of agricultural livestock applications [for genetic engineering] aimed at improving animal productivity; food quality and disease resistance; and environmental sustainability." Productivity of farm animal species can be increased using genetic engineering. Examples include transgenic pigs and sheep that have been genetically altered to express higher levels of growth hormone.

Genetically engineered farm animals can be created to enhance food quality. For example, pigs have been genetically engineered to express the Ä12 fatty acid desaturase gene (from spinach) for higher levels of omega-3, and goats have been genetically engineered to express human lysozyme in their milk. Such advances may add to the nutritional value of animal-based products.

Farm species may be genetically engineered to create disease-resistant animals. Specific examples include conferring immunity to offspring via antibody expression in the milk of the mother; disruption of the virus entry mechanism (which is applicable to diseases such as pseudorabies); resistance to prion diseases; parasite control (especially in sheep); and mastitis resistance (particularly in cattle).

Genetic engineering has also been applied with the aim of reducing agricultural pollution. The best-known example is the Enviropig; a pig that is genetically engineered to produce an enzyme that breaks down dietary phosphorus (phytase), thus limiting the amount of phosphorus released in its manure.

Despite resistance to the commercialization of genetically engineered animals for food production, primarily due to lack of support from the public, a recent debate over genetically engineered AquAdvantage Atlantic salmon may result in these animals being introduced into commercial production.

Effort has also been made to generate genetically engineered farm species such as cows, goats, and sheep that express medically important proteins in their milk. According to Dyck et al, "transgenic animal bioreactors represent a powerful tool to address the growing need for therapeutic recombinant proteins."

In 2006, ATryn® became the first therapeutic protein produced by genetically engineered animals to be approved by the Food and Drug Administration (FDA) of the United States. This product is used as a prophylactic treatment for patients that have hereditary antithrombin deficiency and are undergoing surgical procedures.

RESEARCH ANIMALS

Biomedical applications of genetically engineered animals are numerous, and include understanding of gene function, modeling of human disease to either understand disease mechanisms or to aid drug development, and xenotransplantation. Through the addition, removal, or alteration of genes, scientists can pinpoint what a gene does by observing the biological systems that are affected. While some genetic alterations have no obvious effect, others may produce different phenotypes that can be used by researchers to understand the function of the affected genes.

Genetic engineering has enabled the creation of human disease models that were previously unavailable. Animal models of human disease are valuable resources for understanding how and why a particular disease develops, and what can be done to halt or reverse the process. As a result, efforts have focused on developing new genetically engineered animal models of conditions such as Alzheimer's disease, amyotrophic lateral sclerosis (ALS), Parkinson's disease, and cancer. However, as Wells (13) points out: "these [genetically engineered animal] models do not always accurately reflect the human condition, and care must be taken to understand the limitation of such models." The use of genetically engineered animals has also become routine within the pharmaceutical industry, for drug discovery, drug development, and risk assessment.

As discussed by Rudmann and Durham: "Transgenic and knock out mouse models are extremely useful in drug discovery, especially when defining potential therapeutic targets for modifying immune and inflammatory responses...Specific areas for which [genetically engineered animal models] may be useful are in screening for drug induced immunotoxicity, genotoxicity, and carcinogenicity, and in understanding toxicity related drug metabolizing enzyme systems."

Perhaps the most controversial use of genetically engineered animals in science is to develop the basic research on xenotrans-plantation — that is, the transplant of cells, tissues, or whole organs from animal donors into human recipients.

In relation to organ transplants, scientists have developed a genetically engineered pig with the aim of reducing rejection of pig organs by human recipients. This particular application of genetic engineering is currently at the basic research stage, but it shows great promise in alleviating the long waiting lists for organ transplants, as the number of people needing transplants currently far outweighs the number of donated organs.

However, as a direct result of public consultation, a moratorium is currently in place preventing pig organ transplantation from entering a clinical trial phase until the public is assured that the potential disease transfer from pigs to humans can be satisfactorily managed. According to Health Canada, "xenotransplantation is currently not prohibited in Canada. However, the live cells and organs from animal sources are considered to be therapeutic products (drugs or medical devices)...No clinical trial involving xenotransplantation has yet been approved by Health Canada".

ETHICAL ISSUES OF GENETIC ENGINEERING

Ethical issues, including concerns for animal welfare, can arise at all stages in the generation and life span of an individual genetically engineered animal. The following sections detail some of the issues that have arisen during the peer-driven guidelines development process and associated impact analysis consultations carried out by the CCAC. The CCAC works to an accepted ethic of animal use in science, which includes the principles of the Three Rs (Reduction of animal numbers, Refinement of practices and husbandry to minimize pain and distress, and Replacement of animals with non-animal alternatives wherever possible).

Together the Three Rs aim to minimize any pain and distress experienced by the animals used, and as such, they are considered the principles of humane experimental technique. However, despite the steps taken to minimize pain and distress, there is evidence of public concerns that go beyond the Three Rs and animal welfare regarding the creation and use of genetically engineered animals.

GENETICS OF SCRAPIE RESISTANCE IN SHEEP

Scrapie is an infectious disease of sheep that affects the central nervous system and is always fatal. Upon necropsy, infected animals will have holes or vacuoles in the tissue of the brain. The disease is classified as a transmissible spongiform encephalopathy (TSE). Some TSE's in other species include bovine spongiform encephalopathy (BSE, Mad Cow Disease), Creutzfeldt-Jakob Disease (CJD) in man, new variant CJD (vCJD) in man thought to result from eating meat from BSE-infected cattle, transmissible mink encephalopathy (TSE), chronic wasting disease (CWD) in some U.S., populations of deer and elk, and feline spongiform ecephalopathy (FSE). Federal and State animal health authorities have considered scrapie a high priority disease for elimination for many years, but the relatively recent evidence that ties vCJD in humans to BSE in cattle has resulted in increased attention on scrapie in sheep.

GENETICS OF SCRAPIE

Scrapie is a 100 per cent fatal, degenerative disease affecting the central nervous system of sheep and goats. The disease has been reported in countries throughout the world with few notable exceptions (Australia and New Zealand). The first case of scrapie was discovered in the United States in 1947. The current incidence is 2/10 of 1 per cent or 1 in 500 U.S., sheep. The incidence of scrapie in the U.S., goat population is not currently known.

While bovine spongiform encephalopathy is believed to be caused by the consumption of contaminated feed (meat and bone meal), scrapie is transmitted during lambing when lambs come into contact with infected placenta and birth fluids from infected ewes. Rams can get scrapie, but are not known to transmit scrapie.

Scrapie is NOT caused by genetics, but the genetic make-up (DNA) of an animal determines whether it will get scrapie if it is exposed to the infective agent. In other words, if a genetically susceptible lamb is exposed to a scrapie-infected placenta, it will develop scrapie. It takes from 2 to 5 years after exposure for an animal to show clinical signs of scrapie. If a genetically resistant lamb is exposed to a scrapie-infected placenta, it will not develop scrapie. No resistant genotypes have been identified in goats.

INCIDENCE

Scrapie was first diagnosed in the U.S., in 1947 in a flock of Suffolk sheep in Michigan that had imported sheep from Canada of U.K., origin for several years. From 1947 through July 2001, scrapie had been diagnosed in over 1,000 flocks, and approximately 1,600 individual cases of natural sheep scrapie and 7 individual cases of natural goat scrapie have been found. Through 1992, the breed distribution was as follows: 87 per cent Suffolk, 6 per cent Hampshire, and 7 per cent other breeds and crosses. While 1,600 appears to be a large number of cases of sheep scrapie, it actually is not since it represents an average of only about 30 cases per year. Relative to most other sheep diseases, the incidence of scrapie is quite low.

CLINICAL SIGNS

Early signs of scrapie include subtle changes in behaviour and temperament. These changes may be followed by scratching and rubbing against fixed objects. Other signs are loss of coordination, weight loss despite a good appetite, biting of feet and limbs, and lip smacking. Gait abnormalities such as high-stepping of the forelegs, hopping like a rabbit, and swaying of the back end may be seen. Signs or effects of scrapie usually do not appear until 2 to 5 years after an animal is infected. Sheep may live 1 to 6 months (sometimes longer) after the onset of clinical signs.

Infected animals show no immune response to the infective agent as is the case with viral and bacterial diseases. There is currently no live-animal test to determine if an animal is infected. This is a major area of research, and some promising live-animal tests are in the development and evaluation stages.

The scrapie agent is thought to be spread most often from ewe to offspring and to other lambs in contemporary lambing groups through contact of lambs with the placenta and placental fluids of infected ewes. Lateral transmission from infected rams to ewes and lambs, from infected ewes to other ewes, and from an infected environment to adult animals are not thought to be major routes of transmission; although there are documented cases of scrapie where such routes of transmission are thought to have occurred.

There is not agreement among scientists on the nature of the infectious agent. However, infected animals have large amounts of an abnormal type of prion protein (denoted as PrP^{Sc}), especially in the central nervous system tissues of the brain and spinal cord. Scrapie-free animals have normal prion protein (PrP^{C}) and no PrP^{Sc}. Some feel that PrP^{Sc} itself is the infectious agent while others feel that PrP^{Sc} is the result of another infectious agent. However, scientific evidence increasingly supports the theory of PrP^{Sc} itself as the infectious agent.

PRION PROTEIN

Prion protein is a constituent of normal mammalian cells. Sheep prion protein is a protein of 256 amino acids in length. There is some variation among sheep in the particular amino acids that comprise their prion protein. However, the PrP^{Sc} found in a scrapie-infected sheep is of the same amino acid sequence as the normal prion protein (PrP^C) found in that sheep. The difference between PrP^{Sc} and PrP^C is in its molecular shape. The prion infectious agent theory contends that PrP^{Sc} infects a normal sheep and serves as a template to change the normal shape of PrP^C to the infective shape of PrP^{Sc}.

Of the 256 locations for amino acids along the sheep prion protein molecule, amino acid changes in at least three of these locations (amino acids number 136, 154, and 171) have been shown to confer increased or decreased susceptibility to scrapie. Each particular amino acid is specified in the prion protein gene by a group of three nucleotide bases known as a codon. Therefore, these three important locations are referred to as codon 136, codon 154, and codon 171 on the prion protein gene.

The most common amino acids coded for at these three codons are: alanine (A) or valine (V) at codon 136, arginine (R) or histidine (H) at codon 154, and glutamine (Q) or Arginine (R) at codon 171. Since genes occur in pairs, sheep have two prion protein genes. Each gene can be the same in which case the sheep will produce only one type of prion protein. If the genes are different, then two types of prion protein are produced. Just considering these three codons, a population of sheep can contain individuals with one of 27 different prion protein genotypes.

Table. Possible Genotypes for the Sheep Prion Protein at Codons 136, 154, and 171 and a Scrapie Resistance Index for each Genotype

Codons 136	154	171	Scrapie Resistance Index for U.S. Sheep[a]
AA	RR	QQ	3
AA	RR	QR	9
AA	RR	RR	13
AA	RH	QQ	3
AA	RH	QR	9
AA	RH	RR	13
AA	HH	QQ	3
AA	HH	QR	9
AA	HH	RR	13
AV	RR	QQ	1
AV	RR	QR	7
AV	RR	RR	11
AV	RH	QQ	1
AV	RH	QR	7
AV	RH	RR	11
AV	HH	QQ	1
AV	HH	QR	7
AV	HH	RR	11
VV	RR	QQ	0
VV	RR	QR	6
VV	RR	RR	10
VV	RH	QQ	0

VV	RH	QR	6
VV	RH	RR	10
VV	HH	QQ	0
VV	HH	QR	6
VV	HH	RR	10

Note:

[a]Index is sum of numeric value assigned for each individual codon genotype: codon 136: AA=3, AV=1, VV=0; codon 154: RR=RH=HH=0; codon 171: RR=10, RQ=6, QQ=0. Higher values indicate greater expected resistance.

GENETIC VULNERABILITY OR RESISTANCE

Genetic resistance to scrapie depends not only on the prion genotype of the sheep but also on the strain of scrapie present. Genotypes found to be resistant to one strain of scrapie have been shown to be susceptible to another strain. It appears that there are differences between the U.S., and Europe in the strains of scrapie that are present. The fact that the U.S., is free of certain scrapie strains is good justification for the stringent regulations for importation of sheep and goats into the U.S., from countries with scrapie.

Codon 136: Sheep homozygous for alanine (AA) have been shown to be more resistant to scrapie than sheep homozygous for valine (VV) or heterozygous (AV) in European studies. While a lower incidence of scrapie is anticipated in AA sheep in Europe than in the other genotypes, some AA sheep still have been diagnosed with scrapie. AA sheep are not 100 per cent resistant to scrapie. In the U.S., amino acid changes at codon 136 appear to be less important to scrapie susceptibility than in Europe. This may be due to the different strains of scrapie found in the two regions.

Codon 154: Of the three important codons, amino acid changes at 154 appear to have a slightly less dramatic effect on scrapie susceptibility than do the other two, and the susceptible genotypes are not consistent across studies. In some studies, sheep with the arginine (R) allele have a lowered incidence of scrapie, and in other studies, sheep with the histidine (H) allele have a lowered incidence.

At the present time, it appears that screening sheep on the basis of codon 154 genotype has little value in increasing resistance to scrapie.

Codon 171: Amino acid changes at codon 171 have a large effect on scrapie susceptibility in sheep in both Europe and the U.S. Virtually no sheep homozygous for arginine (RR) have been identified with scrapie. The one exception is a single RR Suffolk in Japan that was diagnosed with scrapie. The frequency of heterozygous (QR) sheep also is very low among scrapie-infected sheep. However, the frequency of sheep homozygous for glutamine (QQ) is very high among scrapie-infected sheep.

A scrapie resistance index for each genotype in U.S., sheep. The higher the number, the greater the expected resistance to scrapie. The index is weighted heavily in favour of RR and QR genotypes at codon 171, less heavily in favour of the AA genotype at codon 136, and neutral with respect to the genotypes at codon 154.

SUSCEPTIBILITY VERSUS SCRAPIE-INFECTED

It must be remembered that a susceptible genotype like QQ at codon 171 does not imply that the animal is scrapie-infected. Scrapie is not a genetic disease. Scrapie is not caused by a particular genotype. An infectious agent causes scrapie. In order to have scrapie, both an infectious agent and a susceptible genotype need to be present. For example, Australia and New Zealand are scrapie-free, but Suffolk, Cheviot, Merino, and Poll Dorset sheep of susceptible genotypes are found in these countries. There is no scrapie in these countries because the scrapie agent is not present.

Likewise, susceptible genotypes are found in many breeds of sheep in the U.S., but almost 90 per cent of the scrapie cases are in the Suffolk breed. This may be due to the major mode of transmission of the disease from an infected ewe to her newborn lambs.

Scrapie was brought into the U.S., in Suffolk sheep. A possible scenario that followed was that infected Suffolk ewes infected their lambs, and infected females from infected ewes continued to pass the disease to their lambs. However, infected Suffolk rams would not tend to infect the ewes they were mated to or their lambs, so infected Suffolk rams mated to ewes of a different breed, as is common in crossbreeding systems in the U.S., would not tend to pass scrapie onto that breed.

While scrapie is primarily a disease of Suffolk sheep in the U.S., the disease is found in the Suffolk, Cheviot, Swaledale, Bleu du Maine, Herdwick, Poll Dorset, Shetland, and Soay breeds in the U.K., Texel breed in The Netherlands, Suffolk and Corriedale breeds in Japan, Romanov and Lacaune breeds in France, Rygja breed in Norway, and Icelandic breed in Iceland. This is not an exhaustive list of sheep breeds or countries with scrapie, but a sample to indicate that scrapie is found in several breeds. A discussion of theories on how scrapie initially becomes established in a breed of sheep would be interesting.

BREEDING FOR SCRAPIE RESISTANCE IN THE U.S.

Selection for resistant prion protein genotypes need only be conducted in breeds where scrapie is found, *i.e.* Suffolks and, perhaps, other blackfaced breeds that have been "improved" by illicit crossing with the Suffolk. With the current strain(s) of scrapie present in the U.S., selection for the arginine (R) allele at codon 171 appears to be sufficient.

A flock where every ewe is of the susceptible genotype QQ can be converted to a flock of QR and RR resistant ewes fairly quickly by simply purchasing and using only RR rams. With a normal ewe replacement rate of 20 per cent per year, over 2/3 of the ewe flock will be of the QR or RR genotypes after five years of RR ram use; even with no selection of ewes or ewe lambs on prion protein genotype. Conversion of the flock to QR and RR ewes will be even faster if, in addition to using RR rams, replacement ewe lambs are preferentially selected from QR or RR dams and if QQ ewes are preferentially culled.

GENOMICS, IMMUNOLOGY AND DISEASES

Considering that 70 per cent of the international fishing grounds are overexploited and that the current volume of captures by fishing efforts has practically reached a maximum level, the increased consumption of fishery related products must be sustained by aquaculture. Indeed, it has recently been estimated that approximately 30 per cent of the fisheries products that are consumed in the world are provided by aquaculture and this aquatic production will increase in the near future. However, mortalities due to disease can occasionally be high in fish farms. Although there are several commercial vaccines that help in the prevention of numerous diseases, further studies are needed for the improvement of existing vaccines and for the development of new vaccines and vaccination strategies. Aquaculture can greatly benefit from the use of molecular and biotechnological tools to identify and characterize genes and regulatory genetic networks of potential use.

In this sense, molecular biology can be applied in aquaculture in the diagnosis of diseases, in the design of new vaccines, in the control of growth and reproduction, and in increasing disease resistance of cultivated species. Moreover the availability of genetic maps of the appropriate density is of high interest for the identification of genomic regions with characters of interest in production by means of assisted marker selection programmes.

This knowledge acquired in fish can be used not only to improve aquaculture production but can also be applied in other industries. In fact, for a long time now, fish have had an impact in Genomics development. Approximately 30 years ago, a popular tropical fish named *Danio rerio* (zebrafish) was chosen as a great candidate for the genetic analysis due to numerous interesting features such as its short life cycle (3 months to reach sexual maturity), relatively high egg clutches throughout all the year, easy culture and external development with transparent embryos. All this combined with large scale mutagenesis experiments conducted at the beginning of the 90s allowed zebrafish studies to fill a gap in the vertebrate development biology because of the ability to help in the study of genes through mutant phenotypes, as in *Drosophila melanogaster* o *Caenorrhabditis elegans*.

Since then great advances on the knowledge and understanding of vertebrate development and human disease have been accrued using this model fish. Fish became an important actor in genomic studies (DNA sequencing and data mining) in 1993, when Sydney Brenner suggested the "pufferfish",

Takifugu rubripes (fugu) as a genome model. On top of its attraction to gourmets in Japan and China fugu has one of the smallest vertebrate genomes. This characteristic already known from its freshwater relative *Tetraodon nigroviridis* in 1968, constitutes a great advantage to access a gene catalogue for a modest cost compared to higher vertebrates. It is important to point out that in recent years available fish genomic resources have increased dramatically, especially for salmonids.

CLONING OF IMMUNE RELATED GENES

A hypothesis that is becoming quite popular among comparative immunologists is that innate immunity of lower vertebrates constitutes an important protection against pathogens for this group of animals. The reason is the relative inefficiency of their acquired immune response due to the evolutionary status and their poikilothermic nature. In fact, innate parameters are considered as relatively independent of temperature and more active at the lower range temperatures at which fish live. However, acquired immunity parameters such as antibody production and lymphocytic activity are more effective at higher temperatures. This translates into a limited antibody repertoire and lower memory and a very slow lymphocyte proliferation in contrast with the almost instantaneous innate response.

Despite of all this, a few non- specific immune genes have been cloned in fish, although this group has grown significantly in the last years. What follows is a review of the "innate and specific" genes characterized up to now in fish (model and commercial species). Most of these studies are based on the cloning and characterization of individual genes, complete sequencing, phylogeny, expression analysis, etc., and recently also the profile expression analysis.

SALMONID FISH

Salmonids, especially rainbow trout (*Oncorhynchus mykiss*), are the group in which the molecular basis of the immune response has been more extensively studied. Among other genes, there have been sequencing and expression studies on interleukins (IL): IL-1β, IL-6, IL-11 and IL-18 and Tumor Necrosis Factor-alpha (TNF-α). Interferon (IFN) and related genes such as Mx have been also studied because of the importance of viral aetiology in diseases and the lack of treatments. In rainbow trout the following genes have been characterized: two Toll Like Receptors (TLRs); IFN regulatory factors (IRFs); Mx proteins; a macrophage activating factor (MAF) similar to IFN-g, expression and gene structure of Inducible Nitric Oxide Synthase (iNOS). Both in rainbow trout and Atlantic salmon (*Salmo salar*) the guanylate binding protein (GBP) induced by IFN-g in mammals has been characterized.

Several antimicrobial peptides (AMPs) have been sequenced and their expression characterized, the same was done with complement factors such as

C4 and molecules type C1, anaphylatoxins C3a, C3a receptor and several chemokines. Transforming Growth Factor-beta (TGF-β) has also been characterized and an Immunoglobulin T (IgT) has been described as well. Concerning IFN-γ inducible chemokines a CXC, CXCL8 type and another with homology with the subgroup CXCL9, CXCL10, CXCL11, apparently with two forms and inducible by viral infections have been identified.

Two chemokines, CK1 and CK2 have been characterized. β1,3-glucan receptors in macrophages have been found in Atlantic salmon. A type C lectin receptor has been sequenced and its expression characterized and a lectin associated to pathogens recognition has been detected in serum. A protein associated to IL-1 receptor has been cloned. Two IgM isotypes and an IgD isotype have been cloned. IFN type I genes and some of the inducible proteins have been cloned and their expression studied, some IFN-g have been identified. A lectin has been partially characterized in Chinook salmon (*Oncorhynchus tshawytscha*) embryos.

NON- SALMONID EUROPEAN FISH

In comparison with those reported in salmonids there are not many characterized genes in non- salmonid fish species. The non- salmonid cloned genes up to date. In sea bream (*Sparus aurata*) several genes induced by interferon (α or γ) such as Mx or IRF-1 have been characterized. Viral infections are able to induce the expression of these factors suggesting an anti-viral function. Sunyer *et al.* have underlined the diversity of functional forms of complement factor 3, as already described in trout. A type F lectin has been characterized. Several factors involved in the immune and inflammatory response have been studied such as TNF, IL-1 or its receptors or TGF. All of these factors are expressed constitutively. In contrary to what is expected with the mammalian perspective, TNF does not increase its expression level after a lipopolysaccharide (LPS) treatment, whilst IL-1 does.

However, TNF shows pro-inflammatory and proliferatory functions when administered *in vivo*, suggesting a similar role as in mammals. There is also evidence of other roles not directly immune such as the regulation of fat tissue in this species. TLR-9 has high expression levels in the immune organs (spleen, head kidney) and in the mucosal and epithelial barriers.

However, its expression did not change in the spleen when fish were infected with bacteria. A second form was found, generated by alternative splicing, although its biological meaning has to be further studied. Recently Major Histocompatibility Complex (MHC) class II alpha chain has been characterized and is constitutively expressed in several tissues and the expression was increased in kidney cells when incubated with bacteria and yeast cells. In addition the CD8a co-receptor has recently been cloned and characterized in sea bream. As in sea bream several genes modulated by interferon α, β and γ have been characterized in turbot (*Psetta maxima*) such as Mx and IRF-1.

Recently, class II α and β MHC and TNF have been cloned. The expression profile in the life cycle has also been studied for hepcidin and T cell receptors.

In sole (*Solea senegalensis*), only the Mx gene has been characterized so far. In sea bass (*Dicentrarchus labrax*) there has been a growing effort in the characterization of immune genes.

Caspase genes such as caspase 3 and 9, cyclooxygenase-2 (COX-2), T cell receptor-β, MHC class II, CD8α, hepcidin and interleukins such as IL-1, IL-10 and IL-12 have all been characterized recently. The light chain of IgM has also been cloned. It is interesting to point out the study on the biological function of IL-1 that has been proposed as an immunostimulant or adjuvant to be used in aquaculture.

NON- SALMONID FISH OUT OF EUROPE

β1,3-glucan receptors have been characterized in catfish (*Ictalurus punctatus*) neutrophils. Also in this species an IgD, Mx and IFN-γ similar sequences have been recently described. In the carp (*Cyprinus carpio*) the α1-anti-protease has been purified and characterized. Two Mannose Binding Protein (MBP) homologues and an MBP associated serine protease have been characterized. Among the different characterized molecules we can quote the following: a chimerical immunoglobulin IgM-IgZ; the protein Nitric Oxide Synthase-2 NOS2; a factor similar to TGF-β-2; a chemokine CXC similar to subgroup CXCL9, CXCL10, CXCL11, a chemokine CC similar to C4-CC, a chemokine receptor CXC and two IL-11 paralogues.

In the herbivorous carp (*Ctenopharyngodon idellus*) the α2-macroglobulin has been purified and characterized. In the Japanese flounder (*Paralichthys olivaceous*) the following immune genes have been cloned: an IgD cDNA; IFN receptor factors; Mx and the chemokines CXC type CXCL8 and two CC chemokines, C6-CC and C4-CC, with a high homology with mammalian CCL11. Three isoforms with homology with mammalian TGF-b-1/4/5, β-2 and β-3 have been described in plaice (*Pleuronectes platessa*). In halibut (*Hippoglossus hippoglossus*) the Mx proteins have also been characterized. In cod (*Gadus morhua*) a soluble and membrane associated IgM form with an unusual alternative splicing and an IgD have been described. In the stripped sea bass (*Morone saxatilis*)a type F lectin has been characterized.

In the cobia (*Rachycentron canadum*), the most important cultured fish in Taiwan, a mannose specific lectin has been cloned and characterized that exhibited a bactericidal and mitogenic activity. The primary structure of a Japanese eel (*Anguilla japonica*) lectin from the epithelial mucus has been characterized. The gene and promoter of viperine (an antiviral protein) from Chinese perch (*Siniperca chuatsi*) have been cloned.

MODEL SPECIES OF PRIMITIVE FISH

The lamprey (*Lampetra fluviatilis*), belonging to the most primitive fish, the agnathans, have been used as a model to study the evolutionary relations with other model fish: fugu, *Tetraodon*, zebrafish, goldfish (*Carassius auratus*) and other species. Although the studies of the immune system are not very abundant, the first fish chemokine is a lamprey CXC that codes a mammal type CXCL8 peptide. A MBP has also been cloned and characterized. Although structurally similar sequences can be found between groups this does not necessarily indicates a functional similarity but rather how conserved the defence related mechanisms can be throughout evolution.

An immunoglobulin isotype has been identified in fugu (*Takifugu rubripes*, fugu), whose structure and expression analysis suggest differences from the IgH locus isotypes previously described; IFN-γ, TLRs, IRFs phagocytic oxidase subunits (oxidase NADPH) and Macrophage Colony-Stimulating Factor (M-CSF) receptor sequences have been also identified. In *Tetraodon nigroviridis*, freshwater fugu, comparative genomics shows an independent expansion in vertebrates of type I IFN receptors and their ligands. Janus kinases (JAKs)-Signal Transducers and Activators of Transcription (STATs) pathway (JAK-STAT pathway) genes have also been characterized.

Several IgZ isotypes have been recently described in zebrafish. A molecular and functional analysis of the following genes have been conducted in zebrafish: IFN, TLRs, several JAK-STAT genes and Mx proteins and their promoters. Sequences of the M-CSF and two chemokines, a CXC similar to mammalian CXCL14 and CXCL12 type have been identified. In goldfish studies have been mostly focused in interferon and associated proteins. TLRs, whose expression increases in activated macrophages, IFN regulatory factors, JAK-STAT genes and antiviral proteins such as Mx, Interferon Stimulated Gene (ISG) and dsRNA Protein Kinase (PKR) have been characterized.

EXPRESSED SEQUENCE TAGS (ESTS)

Immune response understanding is crucial for minimising losses in aquaculture. The use of immunostimulants, vaccines and genetic selection programmes has helped to attain this goal. The use of genomic tools such as ESTs libraries, subtractive hybridizations (SSH) and microarrays have given essential information to identify homologous genes, new genetic functions, expression profiles, routes, candidate genes. EST production, directly from organs related with the function under study or by SSH, has generated multiple databases with hundreds or thousand of genes, which constitute the platform needed for microarray construction.

The developing and application of bioinformatics tools in parallel has permitted an accurate management of a large number of sequences for expression profiles and gene annotation. The developing of microarrays for the simultaneous analysis of the expression of hundreds to thousands of genes is a basic tool for the identification of signatures and candidate genes. The main advantage of the modern techniques in genomics is the large quantity of results produced but this also implies two different risks: the massive production of data in few individuals leads to the choice of models with too high parameters (overparameterization) and obtaining of false positives.

Obtaining consistent results by this strategy demands an appropriate experimental design, a correct application of statistics in data analysis, and recently, the convenience of results modelling using genetic nets has been demonstrated in order to correctly interpret gene expression patterns.

The increase of EST databases has been dramatic in recent years. Salmonids are the group in which this is more apparent. More than 175 cDNA libraries have been constructed from a wide variety of tissues and different developmental stages and more than 300.000 salmonid cDNA sequence reads have been combined from a consortium comprising groups from Canada (Ben Koop and Willie Davidson and the Genomics Research on Atlantic Salmon Project, GRASP; Susan Douglas *et al.* and the Institute for Marine Biosciences, IMB-NRC); France (Yann Guiguen *et al.* and INRA-SCRIBE); Norway (Bjorn Hoyheim *et al.* and the Norwegian School of Veterinary Science, NSVS) and the U.S.A. (Caird Rexroad III and the USDA/ARS National Center for Cool and Cold Water Aquaculture).

These sequences have been assembled into over 40.000 unique contigs. A preliminary microarray of 3.557 cDNAs has been constructed and assessed for its ability to provide new data in the study of cellular and tissue responses to pollutants, diseases and stress, as well as for reproduction and development. On the basis of these results, a larger array of 16.006 genes has been constructed and initial results have shown sensitivity of gene expression patterns to disease challenge, and to small environmental and physiological changes.

BASSMAP, an EU funded project, is focused on sea bass genomics. A Bacterial Artificial Chromosome (BAC) library was constructed with an insert mean of 164 kb and a final cover equivalent to 7 times of total genome. The BAC clones are being used to generate ESTs in order to detect genetic markers such as Single Nucleotide Polymorphisms (SNPs) and microsatellites.

BRIDGEMAP is another EU funded project focused on sea bream genomics, where 1.500 ESTs have been generated with the aim of improving the knowledge of several aspects of the biology of this species. EU has funded and is supporting several research projects that have the common goal of increasing the knowledge of genomics in aspects such as health (AVINSI, AQUAFIRST, IMAQUANIM), welfare (WEALTH, STRESSGENE), reproduction (PROBASS, CRYOCITE, PUBERTIMING), growth and nutrition (FISHCAL, fPPARS). Moreover the network "Marine Genomics Europe" (MGE) has a major goal for the development, use and dissemination of genomic approaches to the study of marine organism biology and marine ecosystems.

One of the goals of the fish and shellfish node has been the production of genomic tools, including cDNA libraries, molecular markers (macro and microsatellites, SNPs), microarrays, etc., integrated in a bioinformatics platform.

These tools will dramatically improve the available genomic resources that will help solving numerous scientific questions. Despite of all these efforts there is a long way in front of us. As an objective data, while the GeneBank number of sequences for rainbow trout is 244.984 and 236.009 for Atlantic salmon, there are only 200 for turbot, 2.448 for sea bream and 24.452 for sea bass. As a comparison there are 833.880 zebrafish ESTs.

IMMUNE RESPONSE AND DISEASE RESISTANCE

Another approach that is being used to characterize genes related with immune response and disease resistance if the study of differential expression of genes using Suppression Subtractive Hybridization (SSH). SSH is a powerful technique that allows the comparison of two mRNA populations, cloning genes that express in one but not in the other. Two pioneer SSH examples are the identification of T cell receptors and the identification of activated genes in *Xenopus laevis* gastrulation. Since then several modifications of the technique have been described.

Alonso and Leong used SSH Chinook salmon and staghorn sculpin (*Hemilepdotus spinosus*) to determine differentially expressed genes against Poly I:C, a potent IFN inducer, finding that the overexpressed genes were quite similar to those described as IFN inducible in mammals. Bayne *et al.* identified rainbow trout immune genes constructing a subtracted library with livers from fish injected with *Vibrio* bacterine. The resulting library consisted in 300-600 bp fragments with 25 genes related with the immune system, of which 15 had not been previously described in salmonid fish and 12 not in any fish at all. The library included acute phase genes such as amyloid A protein, transferrine and precerebeline type protein, complement proteins and lectins, a putative antimicrobial peptide, several membrane receptors, such as TLRs and IL-1 receptor.

O'Farrell *et al.* also used SSH to identify changes in host cells against viral infections (rainbow trout/Viral Hemorrhagic Septicemia, VHS). 24 genes were identified as induced by the virus, most of them were IFN inducible. Zhang *et al.* used SSH to study the model grass carp/Grass Carp Hemorrhagic Virus (GCHV) to conclude that IFN and IFN induced genes are the main tool against virus.

Also, with a viral infection model using Chinese perch and Infectious Spleen and Kidney Necrosis Virus (ISKNV), He *et al.* found genes related with the immune response and apoptosis and with the proteasome degradation pathway of ubiquinated protein that was found also in the SSH library generated by O'Farrell *et al.* and also by Dios *et al*.

Another approach was cDNA libraries generation from primed organs and randomly sequencing of a group of clones. These were successful approaches but with high cost and low efficacy. The same approach was used by Fernandez-Gonzalez *et al.* to detect immune genes in carp experimentally infected with *Icthyophthirius multifilis* sampled at 3 and 72 hours post-infection. In a total of 3.500 ESTs, 82 ortologues with immune relevance, previously described in other organisms, were found. 61 of them have not been previously described in carp. Among them the complete Prostaglandin D2 Synthetase (PGDS), the chemokine SCYA103 and a second βm2 macroglobulin gene were found.

The expression modulation by this parasite was determined by RealTime-PCR (RT-PCR). Tsoi *et al.* used SSH to determine the differential expression of Atlantic salmon genes against the furunculosis etiological agent *Aeromonas salmonicida*. Subtracted cDNA libraries were prepared from three immune organs at two times after infection. A forward and a reverse library were constructed and 200 clones of each library were sequenced, giving a total of 1.778 ESTs that were annotated according functional categories and deposited in GenBank. Numerous genes involved in signalling, innate immunity and other processes were identified.

They include genes that are involved in acute phase and other more novel genes such as tachylectin, hepcidin, genes similar to precerebeline or methyltransferase, a putative protein that binds saxitoxine, etc. A subgroup of genes was studied with more detail by virtual Northern or RT-PCR to verify the differential expression as a result of the infection.

Matejusova *et al.* used SSH to determine *Gyrodactylus salaris* infected Atlantic salmon gene expression profiles. Dios *et al.* used this technique to determine gene profile expression in sea bream brain after nodavirus infection that has the CNS as target. Forward and reverse libraries were generated one day post-infection and the ESTs expressed in infected tissues were catalogued as belonging to stress an immune response. In the reverse library (ESTs expressed in controls when compared with infected tissues) the most expressed genes were ribosomal and mitochondrial.

PHYSICAL PLATFORMS OF MICROARRAYS

Microarrays are physical platforms or chips (usually glass slides) with genes or gene sequences printed, which allow for the simultaneous analysis of the expression of hundred or thousands of genes. The study of these expression profiles in tissues or cell pools constitute the basis for functional genomics. Microarrays can be prepared by printing ESTs, generated from cDNA libraries (subtracted or not) or by short oligos (25-70 bp), designed from the ESTs specific regions such as 3' Untranslated Regions (3'UTRs) or code regions, if alternative splicing studies are the goal. Sometimes cDNA microarrays are prone to the ambiguous EST identification, as they can produce cross-hybridizations among homologous genes and they are not able to distinguish different products of the same gene resulting from differential splicing. In contrast, oligo microarrays avoid some of these problems, but are more expensive. Thus, its application in projects of species involving an important research using arrays could be justified.

Comparative studies between different platforms do not permit a definitive conclusion in this matter. In fact, controversial results between oligo and cDNA platforms have been described. The use of microarrays allows the detection of gene expression profile changes under different experimental conditions such as pathogens, probiotics, developmental stages, contaminants, etc..

The use of microarrays in fish diseases studies was based on human cDNA microarrays. Commercially available human cDNA microarrays were used to compare differential expression in the livers of Atlantic salmon infected with *A. salmonicida* compared to healthy fish. cDNA probes were prepared from total RNA isolated from livers of control salmon and infected salmon by reverse transcription in the presence of 33P-dCTP and independently hybridized to human GENEFILTERS GF211 microarrays.

Of the 4.131 known genes on the microarray, 241 spots gave clearly detectable signals using labelled RNA from the control salmon liver. Of these, 4 spots were consistently found to have a greater than 2-fold increase in infected salmon compared with controls when using the same pair of filters to generate hybridization data from triplicates. These up-regulated genes were ADP/ATP Translocase (AAT2), Na+/K+ ATPase, Acyloxyacyl Hydrolase (AOAH), and Platelet-Derived Growth Factor (PDGF-A). A BlastN search revealed an AAT2 homolog from Atlantic salmon, and a reverse transcriptase polymerase chain reaction assay using primers based on this sequence confirmed its up-regulation (approx. 1.8-fold) during early infection. This work demonstrates the feasibility of using human microarrays to facilitate the discovery of differentially expressed genes in Atlantic salmon.

This was corroborated by Renn *et al.*, which used a microarray prepared with 4.500 genes from a specific cichlid *Astalotilapia burtoni* brain cDNA library. This microarray was used with different fish species and the degree of concordance in expression profiles (number of genes and changes in the levels of expression) was consistent. Rise *et al.* used microarrays constructed with ESTs previously described to study macrophage and haematopoietic kidney gene expression modulated by *Piscirickettsia salmonis* infection. Results were validated by RT-PCR. In infected salmon macrophages, 71 different transcripts were up-regulated and 31 different transcripts were down-regulated. In infected haematopoietic kidney, 30 different transcripts were up-regulated and 39 different transcripts were down-regulated.

Ten antioxidant genes, including glutathione *S*-transferase, glutathione reductase, glutathione peroxidase, and cytochrome b558 and subunits, were up-regulated in infected macrophages but not in infected haematopoietic kidney. Changes in redox status of infected macrophages may allow these cells to tolerate *P. salmonis* infection raising the possibility that the treatment with antioxidants may reduce haematopoietic tissue damage caused by this rickettsial infection. The down-regulation of transcripts involved in adaptive immune responses (*e.g.*, T cell receptor chain and C-C chemokine receptor 7) in infected haematopoietic kidney but not in infected macrophages may contribute to infection-induced kidney tissue damage.

Molecular biomarkers of *P. salmonis* infection, characterized by immune-relevant functional annotations and high fold differences in expression between infected and non- infected samples, may aid in the development of anti-piscirickettsial vaccines and therapeutics. Purcell *et al.* also used this microarray to study Atlantic salmon gene profile against a DNA vector or an Infectious Haematopoietic Necrosis Virus (IHNV) DNA vaccine. Eighty different genes were significantly modulated in the DNA vector group while 910 genes were modulated in the IHNV DNA vaccinated group relative to control group. RT-PCR was used to examine expression of selected immune genes at the Intra Muscular (I.M.) site and in other secondary tissues.

In the localized response (I.M. site), the magnitude of gene expression changes was much greater in the vaccinate group relative to the DNA vector group for the majority of genes analyzed. At secondary systemic sites (*e.g.* gill, kidney and spleen), type I IFN-related genes were up-regulated in only the IHNV DNA vaccinated group. The results presented here suggest that the IHNV DNA vaccine induces up-regulation of the type I IFN system across multiple tissues, which is the functional basis of early anti-viral immunity.

In a similar way Martin *et al.* used this microarray to study the response of Atlantic salmon against an *A. salmonicida* vaccine and found that the greatest increase in expression identified in the array analysis was a liver antibacterial peptide, hepcidin, which was increased 11-fold following the challenge. Roberge *et al.* also used this microarray to determine the levels for gene expression in *Saprolegnia* infected salmon, confirming the importance of non- specific immune response in the resistance against this disease.

MacKenzie *et al.* used another microarray developed by Krasnov *et al.* including 1.380 genes printed in six replicates. They studied the ability of cortisol to directly modulate the transcriptional response of rainbow trout macrophages to the cellular activator LPS. The results indicate that cortisol significantly inhibits the well-described LPS-dependent induction of the expression of TNF-alpha2, a pro-inflammatory cytokine through a complex network of interactions. Gerwick *et al.* used an oligo microarray to study gene profile in the inflammatory process of de *Listonella* (*Vibrio*) *anguillarum* bacterine injected to rainbow trout livers with Freund adjuvant. Microarray analysis determined that individual variability was high probably due to variable resistance to the disease. Li and Waldbieser have also reported the construction and use of a microarray for catfish aimed to the study of innate immune response.

The high density microarray was prepared from oligos based on catfish ESTs. This platform was used to study the gene profile in catfish spleens 2, 4, 8 and 24 hours post-LPS injection. 38 genes were modulated by LPS treatment. The expression of 9 genes determined by RT-PCR was positively correlated with microarray data.

Byon *et al.* determined the antiviral response in Japanese flounder using a microarray. Non-specific immune response genes such as Macrophage Inflammatory Protein 1-a (MIP 1-a) receptor of Natural Killer (NK) and Kupffer cells and Mx1 protein gene were observed to be up-regulated by the VHSV G-protein DNA vaccine at 1 and 3 days post-immunization.

Also, specific immune-related genes including the CD20 receptor, CD8 alpha chain, CD40 and B lymphocyte cell adhesion molecule were also up-regulated during that time. In a later work Byon *et al.* determined that humoral defence-related genes such as complement component C3, complement regulatory plasma proteins, IgM, IgD, MHC class II-associated invariant chain and CD20 receptor were observed to be up-regulated by the VHSg recombinant protein vaccine at 1 or 21 days post-vaccination.

On the other hand, cellular defence-related genes such as CD8 alpha chain, T-cell immune regulator, MIP 1-a and apoptosis-associated protein were not detected. With a second version of this microarray Matsuyama *et al.* determined the immune response of Japanese flounder against *Edwardsiella tarda*. Among the 1.187 analyzed genes, 42 genes were up-regulated during the course of infection either in vaccinated or non-vaccinated fish.

These genes included immune-related genes, such as Matrix Metalloproteinases (MMP) MMP-9, MMP-13, CXC chemokine, CD20 receptor and hepcidin. Some immune-related genes were down-regulated after the *E. tarda* challenge, *i.e.* interferon inducible Mx protein, MHC class II-associated invariant chain, MHC class II alpha and MHC class II beta encoding genes, immunoglobulin light chain precursor, immunoglobulin light chain and IgM. Kurobe *et al.* using cDNA microarray containing 871 unique cDNAs including 91 putative immune-related genes studied the gene expression of *in vitro* grown kidney cells stimulated with mitogens such as Concanavalin A (ConA), Phorbol Myristate Acetate (PMA), LPS or infected with hirame rhabdovirus (HRV).

The number of genes whose expression was increased or decreased by these factors was: 17 by Con A, 139 by PMA, 76 by LPS and 182 by HRV infection. The treatment of ConA for 1 and 6 h affected the expression of only a few of the immune-related genes. PMA down-regulated much more genes from the ones it up-regulated.

Apoptosis-related factors such as c-fos, NGF induced protein IB and NR13 genes were among the genes whose expression was induced by PMA. LPS induced the expression of inflammation-related genes, such as IL-1β, monocyte chemotactic protein 1 and collagenase. The expression of many genes was induced after 3 h HRV infection but some of them were decreased to the basal level after 6 h HRV infection. The expression of some genes of unknown function was induced or reduced by Con A, PMA or LPS or by HRV infection in different time periods. Dios *et al.* studied gene profile expression of nodavirus infected sea bream. cDNA SSH libraries were generated 1 day post-infection.

Some of the genes were included in functional categories of stress and immune system. To check the modulation of expression profiles, samples were taken 1, 3 and 7 days post-infection and hybridized with a macroarray with 765 genes constructed from PCR products with more than 300 bp from the SSH forward (385 genes) and reverse (380) libraries. Significant changes of antiviral genes were detected.

4

Kinds of Diseases

BLASTOMYCOSIS DISEASE

Blastomycosis (also known as "North American blastomycosis," "Blastomycetic dermatitis," and "Gilchrist's disease") is a fungal infection caused by the organism *Blastomyces dermatitidis*. Endemic to portions of North America, blastomycosis causes clinical symptoms similar to histoplasmosis.

CAUSE

B. dermatitidis is asexual form of Ajellomyces dermatitidis. (These can be differentiated on basis of presence or absence of "A Antigen", *B. dermatitidis* being positive for it.) It is a dimorphic fungus.

PATHOGENESIS

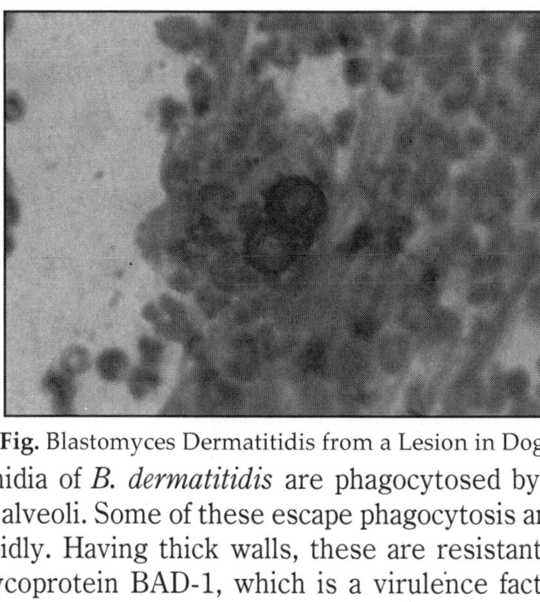

Fig. Blastomyces Dermatitidis from a Lesion in Dog.

Inhaled conidia of *B. dermatitidis* are phagocytosed by neutrophils and macrophages in alveoli. Some of these escape phagocytosis and transform into yeast phase rapidly. Having thick walls, these are resistant to phagocytosis and express glycoprotein BAD-1, which is a virulence factor as well as an epitope. In lung tissue, they multiply and may disseminate through blood and lymphatics to other organs, including the skin, bone, genitourinary tract, and brain. The incubation period is 30 to 100 days, although infection can be asymptomatic.

SIGNS AND SYMPTOMS

Blastomycosis can present in one of the following ways:

- A flu-like illness with fever, chills, arthralgia (joint pain), myalgia (muscle pain), headache, and a non-productive cough which resolves within days.
- An acute illness resembling bacterial pneumonia, with symptoms of high fever, chills, a productive cough, and pleuritic chest pain.
- A chronic illness that mimics tuberculosis or lung cancer, with symptoms of low-grade fever, a productive cough, night sweats, and weight loss.
- A fast, progressive, and severe disease that manifests as ARDS, with fever, shortness of breath, tachypnea, hypoxemia, and diffuse pulmonary infiltrates.
- Skin lesions, usually asymptomatic, can be verrucous (wart-like) or ulcerated with small pustules at the margins.
- Bone lytic lesions can cause bone or joint pain.
- Prostatitis may be asymptomatic or may cause pain on urinating.
- Laryngeal involvement causes hoarseness.
- 40 per cent immunocompromised individuals have CNS involvement and present as brain abscess, epidural abscess or meningitis.

DIAGNOSIS

Once suspected, the diagnosis of blastomycosis can usually be confirmed by demonstration of the characteristic broad based budding organisms in sputum or tissues by KOH prep, cytology, or histology.

Tissue biopsy of skin or other organs may be required in order to diagnose extra-pulmonary disease. Blastomycosis is histologically associated with granulomatous nodules. Commercially available urine antigen testing appears to be quite sensitive in suggesting the diagnosis in cases where the organism is not readily detected. While culture of the organism remains the definitive diagnostic standard, its slow growing nature can lead to delays in treatment of up to several weeks. However, sometimes blood and sputum cultures may not detect blastomycosis; lung biopsy is another option, and results will be shown promptly.

TREATMENT

Itraconazole given orally is the treatment of choice for most forms of the disease. Ketoconazole may also be used. Cure rates are high, and the treatment over a period of months is usually well tolerated. Amphotericin B is considerably more toxic, and is usually reserved for immunocompromised patients who are critically ill and those with central nervous system disease. Patients who cannot tolerate deoxycholate formulation of Amphotericin B can be given lipid formulations. Fluconazole has excellent CNS penetration and is useful where there is CNS involvement after initial treatment with Amphotericin B.

PROGNOSIS

Mortality rate in treated cases:

- 0-2 per cent in treated cases among immunocompetent patients
- 29 per cent in immunocompromised patients
- 40 per cent in the subgroup of patients with AIDS
- 68 per cent in patients presenting as acute respiratory distress syndrome (ARDS)

EPIDEMIOLOGY

In the United States, blastomycosis is endemic in the Mississippi river and Ohio river basins and around the Great Lakes. The annual incidence is less than 1 case per 100,000 people in Mississippi, Louisiana, Kentucky, and Arkansas. The cases are greater in northern states such as Wisconsin, where from 1986 to 1995 there were 1.4 cases per 100,000 people. It also frequently affects hunting dogs in northern Wisconsin and the upper Mississippi and Wisconsin Rivers. In Canada, most cases of blastomycosis occur in Northwestern Ontario, particularly around the Kenora and areas like Killarney (Georgian Bay) Area. The moist, acidic soil in the surrounding woodland harbours the fungus. Blastomycosis is distributed internationally; cases are sometimes reported from Africa.

COCCIDIOIDOMYCOSIS DISEASE

Coccidioidomycosis, commonly known as “Valley fever”, as well as “California fever”, “Desert rheumatism”, and “San Joaquin Valley fever”) is a fungal disease caused by *Coccidioides immitis* or *C. posadasii*. It is endemic in certain parts of Arizona, California, Nevada, New Mexico, Texas, Utah and northern Mexico. *C. immitis* resides in the soil in certain parts of the southwestern United States, northern Mexico, and parts of Central and South America. It is dormant during long dry spells, then develops as a mold with long filaments that break off into airborne spores when the rains come. The spores, known as arthroconidia, are swept into the air by disruption of the soil, such as during construction, farming, or an earthquake. Infection is caused by inhalation of the particles. The disease is not transmitted from person to person. The infection ordinarily resolves leaving the patient with a specific immunity to re-infection.*C. immitis* is a dimorphic saprophytic organism that grows as a mycelium in the soil and produces a spherule form in the host organism.

SIGNS AND SYMPTOMS

Symptomatic infection (40 per cent of cases) usually presents as an influenza-like illness with fever, cough, headaches, rash, myalgia (muscle pain), and arthralgia (joint pain). The rash is maculopapular. Erythema nodosum on lower extremities, and erythema multiforme in necklace-like fashion can occur predominantly in women. Some patients fail to recover and develop chronic pulmonary infection or widespread disseminated infection (affecting meninges, soft tissues, joints, and bone). Severe pulmonary disease may develop in HIV-infected persons.

In order of decreasing risk, people of Filipino, African, Native American, Hispanic, and Asian descent are more susceptible to the disseminated form of the disease than general population. Immunocompromised individuals are more susceptible to the disease.

TYPES

Coccidioidomycosis may be divided into the following types:

- Primary pulmonary coccidioidomycosis
- Disseminated coccidioidomycosis
- Primary cutaneous coccidioidomycosis

COMPLICATIONS

Serious complications include severe pneumonia, lung nodules, and disseminated disease, where the fungus spreads throughout the body. The disseminated form of valley fever can devastate the body, causing skin ulcers, abscesses, bone lesions, severe joint pain, heart inflammation, urinary tract problems, meningitis, and often death.

CAUSE

In soil (and also in agar media), *coccidioides* exists in filament form. It forms hyphae in both horizontal and vertical directions. With time, cells within hyphae degenerate to form alternating barrel-shaped cells (arthroconidia). Arthroconidia are light-weight and carried by air currents. They can be easily inhaled without the person knowing.

On arriving in alveoli, they enlarge in size and internal septations are developed. This structure is called a spherule. Septations develop to form endospores. Rupture of spherules release these endospores, which in turn repeat the cycle and spread the infection locally. Nodules can form in lungs surrounding these spherules.

When these rupture, they release their contents into bronchi, forming thin-walled cavities. These cavities can result in symptoms like characteristic chest pain, hemoptysis and persistent cough. In immunocompromised individuals, infection spreads through blood.

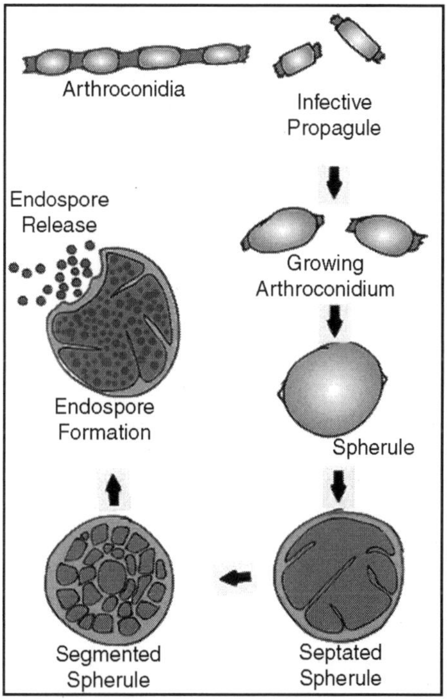

Fig. Life Cycle of Coccidioides

DIAGNOSIS

The fungal infection can be demonstrated by microscopic detection of diagnostic cells in body fluids, exudates, sputum and biopsy-tissue by methods of Papanicolaou or Gomori's methenamine silver staining. These stains can demonstrate spherules and surrounding inflammation. With specific nucleotide primers *C.immitis* DNA can be amplified by PCR. It can also be detected in culture by morphological identification or by using molecular probes that hybridise with *C.immitis* RNA. *C. immitis* and *C. posadasii* cannot be distinguished on cytology or by symptoms; but only by DNA PCR.

An indirect demonstration of fungal infection can be achieved also by serologic analysis detecting fungal antigen or host IgM or IgG antibody produced against the fungus. The available tests include the tube-precipitin (TP) assays, complement fixation assays, and enzyme immunoassays. TP antibody is not found in CSF. TP antibody is specific and is used as confirmatory test, while ELISA is sensitive and thus used for screening. Chest X-Ray usually shows nodules in upper lobes of the lung, usually less than 4 cm in diameter. They rarely calcify. If meninges are affected, CSF will show hypoglycorrhachia (abnormally low glucose levels in CSF), decreased proteins and lymphatic pleocytosis. Rarely, CSF eosinophilia is present. The disease is commonly misdiagnosed as community acquired pneumonia and sometimes as cancer.

TREATMENT

There are no published prospective studies that examine optimal antifungal therapy for coccidioidomycosis. Mild cases often do not require treatment. Oral Fluconazole and intravenous Amphotericin B are used in progressive or disseminated disease, or in which patients are immunocompromised. Alternatively, itraconazole or ketoconazole may be used. Fluconazole is preferred drug for coccidioidal meningitis, due to its penetration into CSF. Intrathecal or intraventricular Amphotericin B therapy is used if infection persists after fluconazole treatment. Itraconazole is used for infection involving bones and joints. Posaconazole and voriconazole have also been used. There is currently no practical preventative measures available for people who live or travel through Valley Fever endemic areas. It is recommended to avoid airborne dust or dirt, though this is not a guaranteed manner of prevention. People in certain occupations may be advised to wear face masks.

EPIDEMIOLOGY

It is confined to the western hemisphere between 40° N and 40° S. Dry soil, especially in the Lower Sonaran Life Zone, is supportive of the organism's growth. In harmony with organism's life history, incidence increases with periods of dryness after a rainy season; this phenomenon, termed "grow and blow", refers to growth of the fungus in wet weather, producing spores which are spread by the wind during succeeding dry weather.

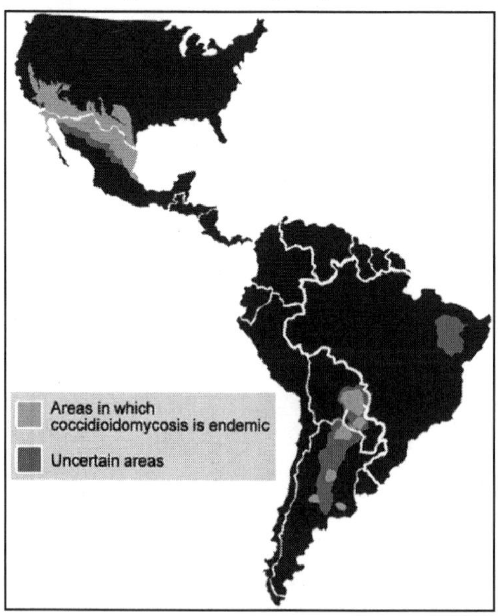

Fig. The Geographic Distribution of Coccidioidomycosis.

Besides humans, dogs, and cats, the fungus can be shown to infect most mammals, even if they do not get sick from it very often. Species in which

Valley Fever has been found include livestock such as cattle and horses; llamas; marine mammals, including sea otter; zoo animals such as monkeys and apes, kangaroos, tigers, etc; and wildlife endemic to the geographic area such as cougar, skunk, and javelina.

BIOLOGICAL WARFARE

C. immitis was investigated by the United States during the 1950s and 1960s as a potential biological weapon. The Cash strain received the military symbol OC, and original hopes were for its use as an incapacitant. As medical epidemiology later made clear, OC would have lethal effects on several segments of the population, so it was later considered a lethal agent. It was never standardised, and beyond a few field trials, it was never weaponised. Most military work on OC was on vaccines by the mid-1960s. It is still on the CDC's list of select agents, however.

CRYPTOCOCCOSIS DISEASE

Cryptococcosis, or cryptococcal disease, is a potentially fatal fungal disease. It is caused by one of two species; *Cryptococcus neo-formans* and *Cryptococcus gattii*. These were all previously thought to be subspecies of *C. neo-formans*, but have now been identified as distinct species.

Cryptococcosis is believed to be acquired by inhalation of the infectious propagule from the environment. Although the exact nature of the infectious propagule is unknown, the leading hypothesis is the basidiospore created through sexual or asexual reproduction.

CAUSE

Cryptococcosis is a defining opportunistic infection for AIDS. Other conditions which pose an increased risk include certain lymphomas (*e.g.* Hodgkin's lymphoma), sarcoidosis, liver cirrhosis and patients on long-term corticosteroid therapy.

Distribution is worldwide in soil. The prevalence of cryptococcosis has been increasing over the past 20 years for many reasons, including the increase in incidence of AIDS and the expanded use of immunosuppressive drugs.

In humans, C. neo-formans causes three types of infections:

- Wound or cutaneous cryptococcosis
- Pulmonary cryptococcosis, and
- Cryptococcal meningitis.

Cryptococcal meningitis (infection of the meninges, the tissue covering the brain) is believed to result from dissemination of the fungus from either an observed or unappreciated pulmonary infection. Often there is also silent dissemination throughout the brain when meningitis is present. *Cryptococcus gattii* causes infections in immunocompetent people (those having a functioning immune system), but *C. neo-formans v. grubii*, and *v. neo-formans* usually only cause clinically evident infections in persons who have some form of defect in their immune systems (immunocompromised persons). People who have defects in their cell-mediated immunity, for example, people with AIDS, are especially susceptible to disseminated cryptococcosis. Cryptococcosis is often fatal, even if treated. The 10-week survival averages near 70 per cent with optimal therapy.

Although the most common presentation of cryptococcosis is of *C. neoformans* infection in an immunocompromised person (such as persons living with AIDS), the *C. gattii* is being increasingly recognised as a pathogen in presumptively immunocompetent hosts, especially in Canada and Australia. This may be due to rare exposure and high pathogenicity, or to unrecognised isolated defects in immunity, specific for this organism.

DIAGNOSIS

Dependent on the infectious syndrome, symptoms include fever, fatigue, chest pain, dry cough, swelling of abdomen, headache, blurred vision and confusion. Detection of cryptococcal antigen (capsular material) by culture of CSF, sputum and urine provides definitive diagnosis. Blood cultures may be positive in heavy infections. India ink of the CSF is a traditional microscopic method of diagnosis, although the sensitivity is poor in early infection, and may miss up to >=20 per cent of patients with culture-positive cryptococcal meningitis. Cryptococcosis can rarely occur in the immunocompetent person without HIV, when it usually goes undiagnosed. Less than 250 cases in all are reported in the medical literature, the majority diagnosed postmortem.

TREATMENT

Treatment options in non-AIDS patients who have reduced immune-system function is not well studied. Intravenous Amphotericin B combined with oral flucytosine may be effective. Every attempt should be made to reduce the amount of immunosuppressive medication until the infection is resolved.

Persons living with AIDS often have a greater burden of disease and higher mortality (30-70 per cent at 10-weeks), but recommended therapy is with Amphotericin B (0.7-1.0 mg/kg/day) and flucytosine. Where flucytosine (5FC) is not available, fluconazole 800-1200mg/day should be used adjunctively with amphotericin. Amphotericin-based induction therapy has much greater microbiologic activity than fluconazole monotherapy with 30 per cent better survival at 10-weeks. Based on a systematic review of existing data, the most cost-effective induction treatment in resource-limited settings appears to be one week of amphotericin B 1mg/kg/day coupled with high dose fluconazole 1200mg/day. After initial treatment as above, oral fluconazole can be used. The decision on when to start treatment for HIV is not yet settled, although one small, under-powered trial (when restricting to a "per protocol analysis") suggested that delaying the start of treatment for several weeks may be beneficial in avoiding deaths from Immune reconstitution inflammatory syndrome (IRIS). A larger multisite trial, the Cryptococcal Optimal ART Timing (COAT) Trial, will be completed in spring 2013.

PREVENTION OF CRYPTOCOCCOSIS

Cryptococcosis is a very subacute infection with a prolonged subclinical phase lasting weeks to months in persons with HIV/AIDS before the onset of symptomatic meningitis. In Sub-Saharan Africa, the prevalence rates of detectable cryptococcal antigen in perpheral blood is often 4-12 per cent in persons with CD4 counts lower than 100 cells/mcL. Cryptococcal antigen screen and pre-emptive treatment with fluconazole is cost saving to the health care system by avoiding cryptococcal meningitis.

The World Health Organisation recommends cryptococcal antigen screening in HIV-infected persons entering care with CD4<100 cells/ìL. This undetected subclinical cryptococcal (if not pre-emptively treated with anti-fungal therapy) will often go on to develop cryptococcal meningitis, despite receiving HIV therapy. Cryptococcosis accounts for 20-25 per cent of the mortality after initiating HIV therapy in Africa.

IRIS IN IMMUNOCOMPETENT HOSTS

The immune reconstitution inflammatory syndrome (IRIS) has been described in immunocompetent hosts who have meningitis caused by *C. gattii* and *C. grubii*. Several weeks or even months into appropriate treatment, there is a sudden onset deterioration with worsening meningitis symptoms and progression or development of new neurological symptoms. IRIS is however much more common in immunocompromised hosts. Magnetic resonance imaging shows increase in the size of brain lesions, and CSF abnormalities (white cell count, protein, glucose) increase. CSF culture is sterile, and there is no increase in CSF cryptococcal antigen titre.

The increasing inflammation can cause brain injury or be fatal. The mechanism behind IRIS in cryptococcal meningitis is primarily immunologic. With reversal of immunosuppression, there is paradoxical increased inflammation as the recovering immune system recognises the fungus. In severe IRIS cases, treatment with systemic corticosteroids has been utilised.

IN OTHER ANIMALS

Cryptococcosis is also seen in cats and occasionally dogs. It is the most common deep fungal disease in cats, usually leading to chronic infection of the nose and sinuses, and skin ulcers. Cats may develop a bump over the bridge of the nose from local tissue inflammation. It can be associated with FeLV infection in cats. Cryptococcosis is most common in dogs and cats but cattle, sheep, goats, horses, wild animals and birds can also be infected. Soil, fowl manure and pigeon droppings are among the sources of infection.

EUMYCETOMA DISEASE

Eumycetoma is a chronic, specific, granulomatous, fungal disease. It mainly affects the foot; and *Mycetoma pedis* is also known as Madura foot. This infection is endemic in Africa, India, and Central and South America.

CAUSES AND PRESENTATION

Eumycetoma usually involves the subcutaneous tissue after a traumatic inoculation of the causative organism. Swelling and formation of sinus tracts characterise mycetoma. The sinuses usually discharge purulent and seropurulent exudate containing grains. It may spread to involve the skin and the deep structures resulting in destruction, deformity and loss of function; very occasionally it could be fatal.

EPIDEMIOLOGY

The true incidence and the geographical distribution of mycetoma throughout the world is not exactly known due to the nature of the disease which is usually painless, slowly progressive which may lead to the late presentation of the majority of patients. Mycetoma has a worldwide distribution but this is extremely uneven. It is endemic in tropical and subtropical regions. The African continent seems to have the highest prevalence. It is found in what is known as the mycetoma belt stretching between the latitudes of 15 south and 30 north. The belt includes Sudan, Somalia, Senegal, India, Yemen, Mexico, Venezuela, Colombia, Argentina and others.

Mycetoma infection, Madura foot or maduromycosis, was originally described in Sanskrit in the Vedic texts from India. The first English language accounts occurred much later in the area of Madras (aka Chennai). The geographical distribution of the individual mycetoma organism shows considerable variations, which can be convincingly explained on an environmental basis. Areas where mycetoma prevails are relatively arid zones with a short rainy season with a relative humidity. The organisms are usually present in the soil in the form of grains. The infecting agent is implanted into the host tissue through a breach in the skin produced by trauma caused by sharp objects such as thorn pricks, stone or splinters.

PATHOGENESIS

The disease is usually acquired while performing agricultural work, and it generally afflicts men between 20 and 40 years old. The disease is acquired by contacting grains of fungal spores that have been discharged onto the soil. Infection usually involves an open area or break in the skin. *Pseudoallescheria boydii* is one of many fungi spp. that causes the fungal form of madura foot. The disease is characterised by a yogurt-like discharge upon maturation of the infection. Hematogenous or lymphatic spread is uncommon. Infections normally start in the foot or hand and travel up the leg or arm.

Eumycetoma may be one of several varieties, depending upon colour of the granulous discharge:

- Red
  - *Actinomadura pelletieri*
- White or yellow
  - *Acremonium strictum*
  - *Actinomadura madurae*
  - *Aspergillus nidulans*
  - *Noetestudina rosatii*
  - *Phaeoacremonium krajdenii*
  - *Pseudallescheria boydii*
- Black
  - *Aspergillus terreus*
  - *Curvularia lunata*
  - *Cladophialophora bantiana*
  - *Exophiala jeanselmei*
  - *Leptosphaeria senegalensis*
  - *Leptosphaeria tompkinsii*
  - *Madurella grisea*
  - *Madurella mycetomatis*
  - *Pyrenochaeta romeroi*

The bacterial mycetoma species *Nocardia* (including *Nocardia asteroides* and *Nocardia brasiliensis*) produces yellow discharge, and *Streptomyces* (including *Streptomcyes somaliensis*) produce yellow or red discharge. These are not causes of Eumycetoma.Botryomycosis; also known as bacterial pseudomycosis is a rare chronic granulomatous bacterial infection that affects the skin, and sometimes the viscera.Staphylococcus aureus is usually the organism that causes the infection, however it can also be caused by Pseudomonas aeruginosa or several other species of bacteria.

The further course of the sinuses differs according to the mycetoma present. In the black variety, the infection spreads mainly subcutaneously. In the red and yellow varieties deep spread occurs early, and muscle and underlying bones become infiltrated, but unexpectedly, nerves and tendons are highly resistant to invasion.

DIAGNOSIS

Diagnosis of mycetoma is usually accomplished by radiology, ultrasound or by fine needle aspiration of the fluid within an afflicted area of the body. It depends upon isolating the causative organism along with a knowledge of local endemic infection.

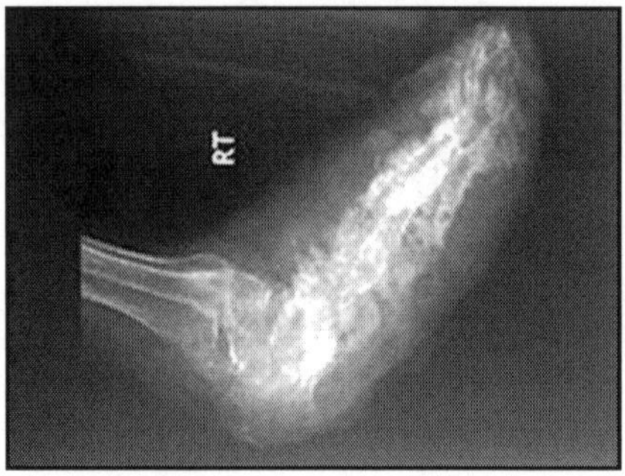

Fig. Madura Foot X-Ray

Differential Diagnosis

Following is the differential diagnosis:

1. Tuberculous ulcer
2. Kaposi's sarcoma
3. Tropical ulcer

TREATMENT

There are several clinical treatments available for this disease. They include surgery, ketoconazole, voriconazole, itraconazole and amputation of the affected limb. There is no vaccine for mycetoma. Scientists at such institutions as The Mycetoma Research Center, the University of Khartoum in Sudan are working on a cure.

HISTOPLASMOSIS DISEASE

Histoplasmosis (also known as "Cave disease," "Darling's disease," "Ohio valley disease," "Reticuloendotheliosis," "Spelunker's Lung" and Caver's disease) is a disease caused by the fungus *Histoplasma capsulatum*. Symptoms of this infection vary greatly, but the disease primarily affects the lungs. Occasionally, other organs are affected; this is called disseminated histoplasmosis, and it can be fatal if left untreated. Histoplasmosis is common among AIDS patients because of their suppressed immunity. In immunocompetent individuals, past infection results in partial protection against ill effects if reinfected.

Histoplasmosis is generally contracted from contact to microscopic fungi borne from decomposing biological fluids, most notably human excretions like urine, vomit, and feces. Cases of histoplasmosis have declined acutely since the Industrial Revolution as quality of life improved dramatically and humans were no longer living in their own squalor. Unfortunately, it is still a major killer in third world countries and can be contracted easily in the first world by living among the aforementioned human fluids.

SYMPTOMS

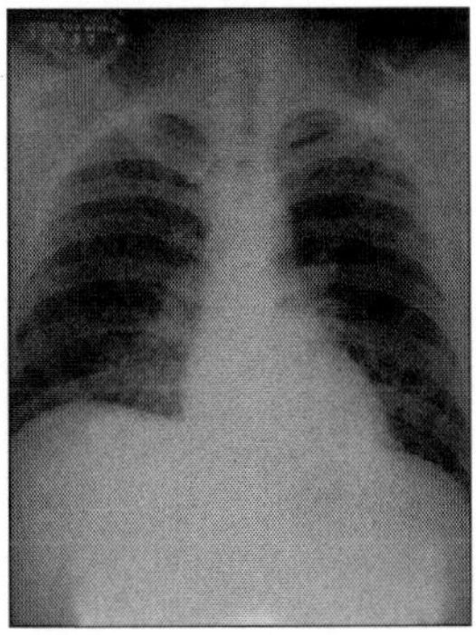

Fig. Chest X-ray of a Patient with Acute Pulmonary Histoplasmosis.

If symptoms of histoplasmosis infection occur, they will start within 3 to 17 days after exposure; the average is 12–14 days. Most affected individuals have clinically silent manifestations and show no apparent ill effects. The acute phase of histoplasmosis is characterised by non-specific respiratory symptoms, often cough or flu-like.

Chest X-ray findings are normal in 40–70 per cent of cases. Chronic histoplasmosis cases can resemble tuberculosis; disseminated histoplasmosis affects multiple organ systems and is fatal unless treated.

While histoplasmosis is the most common cause of mediastinitis, this remains a relatively rare disease. Severe infections can cause hepatosplenomegaly, lymphadenopathy, and adrenal enlargement. Lesions have a tendency to calcify as they heal. Ocular histoplasmosis damages the retina of the eyes. Scar tissue is left on the retina which can experience leakage, resulting in a loss of vision not unlike macular degeneration.

TYPES

Histoplasmosis may be divided into the following types:

- Primary pulmonary histoplasmosis
- Progressive disseminated histoplasmosis
- Primary cutaneous histoplasmosis
- African histoplasmosis

EPIDEMIOLOGY

"Histoplamosis capsulatum" is found throughout the world. It is endemic in certain areas of the United States, particularly in states bordering the Ohio River valley and the lower Mississippi River. The humidity and acidity patterns of soil are associated with endemicity. Bird and bat droppings in soil promote growth of *Histoplasma*. Contact with such soil aerosolises the microconidia, which can infect humans. It is also common in caves in southern and East Africa. Positive histoplasmin skin tests occur in as many as 90 per cent of the people living in areas where *H. capsulatum* is common, such as the eastern and central United States.

DISEASE MECHANISM

H. capsulatum grows in soil and material contaminated with bird or bat droppings (guano). The fungus has been found in poultry house litter, caves, areas harbouring bats, and in bird roosts (particularly those of starlings). The fungus is thermally dimorphic: in the environment it grows as a brownish mycelium, and at body temperature (37°C in humans) it morphs into a yeast. Histoplasmosis is not contagious, but is contracted by inhalation of the spores from disturbed soil or guano.

The inoculum is represented principally by microconidia. These are inhaled and reach the alveoli. In the alveoli, macrophages ingest these microconidia. They survive inside the phagosome. As the fungus is thermally dimorphic, these microconidia are transformed into yeast. They grow and multiply inside the phagosome.

The macrophages travel in lymphatic circulation and spreading the disease to different organs. Cell mediated immunity for histoplasmosis develops within 2 weeks. If the patient has strong cellular immunity, macrophages, epithelial cells and lymphocyte surround the organism and contain them, and eventually calcify. In immunocompromised individuals, the organisms disseminate to different organs such as bone, spleen, liver, adrenal glands, mucocutaneous membranes, resulting in Progressive disseminated histoplasmosis. Chronic lung disease can manifest.

DIAGNOSIS

Clinically, there was a wide spectrum of disease manifestation. In this meaning, doctor's integration is very important.

More severe forms includes:

- The chronic pulmonary form, often occurring in the presence of underlying pulmonary disease; and
- A disseminated form, which is characterised by the progressive spread of infection to extra-pulmonary sites.

Oral manifestations have been reported as the main complaint of the disseminated forms, leading the patient to look for assistance, whereas pulmonary symptoms may be mild or even misinterpreted as flu. Histoplasmosis can be diagnosed by samples containing the fungus taken from sputum (via Bronchoalveolar lavage), blood, or infected organs. It can also be diagnosed by detection of antigens in blood or urine samples by ELISA or PCR. Antigens can cross-react with antigens of African histoplasmosis (caused by Histoplasma duboisii), blastomycosis, coccidioidomycosis, paracoccidioidomycosis, and Penicillium marneffei infection. It can also be diagnosed by a test for antibodies against *Histoplasma* in the blood. *Histoplasma* skin tests indicate whether a person has been exposed, but do not indicate whether they have the disease. Formal histoplasmosis diagnoses are often confirmed only by culturing the fungus directly.

Cutaneous manifestations of disseminated disease are diverse and often present as a non-descript rash with systemic complaints. Diagnosis is best established by urine antigen testing. Blood cultures may take up to 6 weeks for diagnostic growth to occur and serum antigen testing often comes back with a false negative before 4 weeks of disseminated infection.

TREATMENT

In the majority of immunocompetent individuals, histoplasmosis resolves without any treatment. Antifungal medications are used to treat severe cases of acute histoplasmosis and all cases of chronic and disseminated disease. Typical treatment of severe disease first involves treatment with amphotericin B, followed by oral itraconazole.

Liposomal preparations of amphotericin B are more effective than deoxycholate preparations. Deoxycholate preparation is preferred in patients who might be at risk of nephrotoxicity. Individuals taking amphotericin B are monitored for renal function. Treatment with itraconazole will need to continue for at least a year in severe cases, while in acute pulmonary histoplasmosis, 6 to 12 weeks treatment is sufficient. Alternatives to itraconazole are posaconazole, voriconazole, and fluconazole. Individuals taking itraconazole are monitored for hepatic function.

COMPLICATIONS

In absence of proper treatment and especially in immunocompromised individuals, complications can arise. These include recurrent pneumonia, respiratory failure, fibrosing mediastinitis, superior vena cava syndrome, pulmonary vessel obstruction, progressive fibrosis of lymph nodes.

Fibrosing mediastinitis is a serious complication and can be fatal. Smokers with structural lung disease have higher probability of developing chronic cavitary histoplasmosis. After healing of lesions, hard calcified lymph nodes can erode the walls of airway causing hemoptysis.

PREVENTION

It is not practical to test or decontaminate most sites that may be contaminated with *H. capsulatum*, but the following sources list environments where histoplasmosis is common, and precautions to reduce a person's risk of exposure, in the three parts of the world where the disease is prevalent. Precautions common to all geographical locations would be to avoid accumulations of bird or bat droppings.

The US National Institute for Occupational Safety and Health (NIOSH) provides a document, which contains information on work practices and personal protective equipment that may reduce the risk of infection. This document is available in English and Spanish.

- Authors at the University of Nigeria have published a review which includes information on locations in which histoplasmosis has been found in Africa (in chicken runs, bats and the caves bats infest, and in soil), and a thorough reference list including English, French, and Spanish language references.

HISTORY

Histoplasma was discovered in 1905 by Darling, but was only discovered to be a widespread infection in the 1930s. Before then, many cases of histoplasmosis were mistakenly attributed to tuberculosis, and patients were mistakenly admitted to tuberculosis sanatoriums. Some patients contracted tuberculosis in these sanatoriums.

PARACOCCIDIOIDOMYCOSIS DISEASE

Paracoccidioidomycosis (also known as "Brazilian blastomycosis," "South American blastomycosis,", "Lutz-Splendore-de Almeida disease" and "Paracoccidioidal granuloma":320) is a mycosis caused by the fungus Paracoccidioides brasiliensis. Sometimes called South American blastomycosis, paracoccidioidomycosis is caused by a different fungus than that which causes blastomycosis.

AGENT

P. brasiliensis is a thermally dimorphic fungus distributed in Brazil and South America. The habitat of the infectious agent is not known but appears to be aquatic. In biopsies the fungus appears as a polygemulating yeast with a pilot's wheel-like appearance.

DISEASE

Paracoccidioidomycosis is a systemic mycosis caused by the dimorphic fungus Paracoccidioides. It frequently involves mucous membranes, lymph nodes, bone and lungs. Unlike other systemic mycoses, it can cause disease in immunocompetent hosts, although immunosuppression increases the aggressiveness of the fungus.

Also uniquely, it rarely causes disease in fertile-age women, probably due to a protective effect of estradiol. Primary infection is thought to be autolimited and almost asymptomatic as histoplasmosis or coccidioidomycosis (Valley Fever). In young people, there is a progressive form of the disease (akin of tuberculous septicemia in tuberculous priminfection) with high prostrating fever, generalised lymphadenopathy and pulmonary involvement with milliary lesions. This juvenile form has a more severe prognosis even with treatment. The most common form is the so-called adult form of paracoccidioidomycosis that is almost certainly a reactivation of the disease.

Painful lesions with a violaceous hue in lips and oral mucosa are common as is cervical lymphadenitis teeming with polygemulating yeasts in the biopsy. In this form, differential diagnosis must be made with mucocutaneous leishmaniasis, yaws and TB.

Pulmonary involvement is also common, it starts as lobar pneumonia or pleurisy but without remission at ninth day; the patient remains febrile, coughs, loses weight and the X-rays reveal milliary shadows throughout lung fields. Other organs can be involved, like bones, meninges, arteries and spleen but this is very rare. Diagnosis is made with a biopsy of affected tissue, this shows the characteristic helm-shaped yeasts and culture shows the agent. Serology is also used in endemic areas.

TREATMENT

Sulfonamides are the traditional remedies to paracoccidiodomycosis. They were introduced by Oliveira Ribeiro and used for more than fifty years with good results. The most used sulfa drugs in this infection are sulfadimethoxime, sulfadiazine and co-trimoxazole.

This treatment is generally safe but several adverse effects can appear, the most severe of which are the Stevens Johnson Syndrome and agranulocytosis. Similarly to tuberculosis treatment, it must be continued for up to 3 years to eradicate the fungus, and relapse and treatment failures aren't unusual.

Antifungal drugs like amphotericin B or itraconazole and ketoconazole are more effective in clearing the infection but limited by their cost when compared with sulfonamides. During therapy fibrosis can appear and a surgery be needed to correct this. Other possible complication is Addisonian crisis. The death rate is approximately ten percent.

DISCOVERY

Lutz-Splendore-de Almeida disease is named for the physicians Adolfo Lutz, Alfonso Splendore, and Floriano Paulo de Almeida, who first characterised the disease in Brazil in the early 20th century.

SPOROTRICHOSIS DISEASE IN ANIMALS

Sporotrichosis (also known as "Rose gardener's disease") is a disease caused by the infection of the fungus *Sporothrix schenckii*. This fungal disease usually affects the skin, although other rare forms can affect the lungs, joints, bones, and even the brain. Because roses can spread the disease, it is one of a few diseases referred to as *rose-thorn* or *rose-gardeners' disease*. Because *S. schencki* is naturally found in soil, hay, sphagnum moss, and plants, it usually affects farmers, gardeners, and agricultural workers.

It enters through small cuts and abrasions in the skin to cause the infection. In case of sporotrichosis affecting the lungs, the fungal spores enter through the respiratory pathways. Sporotrichosis can also be acquired from handling cats with the disease; it is an occupational hazard for veterinarians. Sporotrichosis progresses slowly - the first symptoms may appear 1 to 12 weeks (average 3 weeks) after the initial exposure to the fungus. Serious complications can also develop in patients who have a compromised immune system.

FORMS AND SYMPTOMS OF SPOROTRICHOSIS

- *Cutaneous or skin sporotrichosis*: This is the most common form of this disease. Symptoms of this form include nodular lesions or bumps in the skin, at the point of entry and also along lymph nodes and vessels.
- The lesion starts off small and painless, and ranges in colour from pink to purple.
- Left untreated, the lesion becomes larger and look similar to a boil and more lesions will appear, until a chronic ulcer develops. Usually, cutaneous sporotrichosis lesions occur in the finger, hand, and arm.
- *Pulmonary sporotrichosis*: This rare form of the disease occur when *S. schenckii* spores are inhaled. Symptoms of pulmonary sporotrichosis include productive coughing, nodules and cavitations of the lungs, fibrosis, and swollen hilar lymph nodes. Patients with this form of sporotrichosis are susceptible to developing tuberculosis and pneumonia.
- *Disseminated sporotrichosis*: When the infection spreads from the primary site to secondary sites in the body, the disease develops into a rare and critical form called disseminated sporotrichosis. The infection can spread to joints and bones (called *osteoarticular sporotrichosis*) as well as the central nervous system and the brain (called *sporotrichosis meningitis*). The symptoms of disseminated sporotrichosis include weight loss, anorexia, and appearance of bony lesions.

SPOROTRICHOSIS IN ANIMALS

Sporotrichosis can be diagnosed in domestic and wild mammals. In veterinary medicine it is most frequently seen in cats and horses. Cats have a particularly severe form of cutaneous sporotrichosis and also can serve as a source of zoonotic infection to persons who handle them and are exposed to exudate from skin lesions.

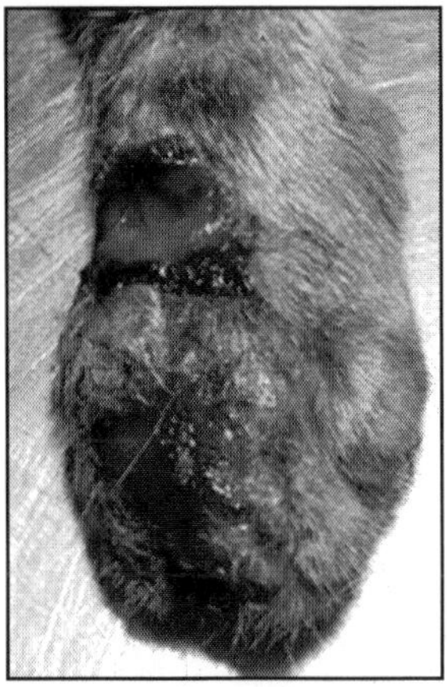

Fig. Ulcerative Skin Disease in a Cat with Sporotrichosis; a Cat with this Disease must be Handled with Caution as this form can be Contagious to other Animals and to Humans

DIAGNOSIS

Sporotrichosis is a chronic disease with slow progression and often subtle symptoms. It is difficult to diagnose, as many other diseases share similar symptoms and therefore must be ruled out. Patients with sporotrichosis will have antibody against the fungus S. schenckii, however, due to variability in sensitivity and specificity, it may not be a reliable diagnosis for this disease. The confirming diagnosis remains culturing the fungus from the skin, sputum, synovial fluid, and cerebrospinal fluid. Cats with sporotrichosis are unique in that the exudate from their lesions may contain numerous organisms. This makes cytological evaluation of exudate a valuable diagnostic tool in this species. Exudate is pyogranulomatous and phagocytic cells may be packed with yeast forms. These are variable in size, but many are cigar-shaped.

PREVENTION

The majority of sporotrichosis cases occur when the fungus is introduced through a cut or puncture in the skin while handling vegetation containing the fungal spores. Prevention of this disease includes wearing long sleeves and gloves while working with soil, hay bales, rose bushes, pine seedlings, and sphagnum moss.

TREATMENT

Treatment of sporotrichosis depends on the severity and location of the disease.

The following are treatment options for this condition:

- *Saturated potassium iodide solution*: Although its mechanism is unknown, application of potassium iodide in droplet form can cure cutaneous sporotrichosis. This usually requires 3 to 6 months of treatment.

- *Itraconazole (Sporanox) and fluconazole*: These are antifungal drugs. Itraconazole is currently the drug of choice and is significantly more effective than fluconazole. Fluconazole should be reserved for patients who cannot tolerate itraconazole.
- *Amphotericin B*: This antifungal medication is delivered intravenously. Many patients, however, cannot tolerate Amphotericin B due to its potential side effects of fever, nausea, and vomiting.

Lipid formulations of amphotericin B are usually recommended instead of amphotericin B deoxycholate because of a better adverse-effect profile. Amphotericin B can be used for severe infection during pregnancy. For children with disseminated or severe disease, amphotericin B deoxycholate can be used initially, followed by itraconazole.

In case of sporotrichosis meningitis, the patient may be given a combination of Amphotericin B and 5-fluorocytosine/Flucytosine:

- *Newer triazoles*: Several studies have shown that posaconazole has *in vitro* activity similar to that of amphotericin B and itraconazole; therefore, it shows promise as an alternative therapy. However, voriconazole susceptibility varies. Because the correlation between *in vitro* data and clinical response has not been demonstrated, there is insufficient evidence to recommend either posaconazole or voriconazole for treatment of sporotrichosis at this time.
- *Surgery*: In cases of bone infection and cavitatory nodules in the lungs, surgery may be necessary.

COMPLICATIONS

Sporotrichosis can develop into other diseases, including cellulitis, sarcoidosis, and tuberculosis.

5

Bacteriology in Animals

THE ORIGIN OF LIFE

When life arose on Earth about 4 billion years ago, the first types of cells to evolve were procaryotic cells. For approximately 2 billion years, procaryotic-type cells were the only form of life on Earth. The oldest known sedimentary rocks, from Greenland, are about 3.8 billion years old. The oldest known fossils are procaryotic cells, 3.5 billion years in age, found in Western Australia and South Africa. The natüre of these fossils, and the chemical composition of the rocks in which they are found, indicates that lithotrophic and fermentative modes of metabolism were the first to evolve in early procaryotes. Photosynthesis developed in bacteria a bit later, at least 3 billion years ago. Anoxygenic photosynthesis (bacterial photosynthesis, which is anaerobic and does not produce O_2) preceded oxygenic photosynthesis (plant-type photosynthesis, which yields O_2). However, oxygenic photosynthesis also arose in procaryotes, specifically in the cyanobacteria, which existed millions of years before the evolution of green algae and plants. Larger, more complicated eucaryotic cells did not appear until much later, between 1.5 and 2 billion years ago. The archaea and bacteria differ fundamentally in their structure from eucaryotic cells, which always contain a membrane-enclosed nucleus, multiple chromosomes, and various other membranous organelles such as mitochondria, chloroplasts, the golgi apparatus, vacuoles, etc. Unlike plants and animals, archaea and bacteria are unicellular organisms that do not develop or differentiate into multicellular forms. Some bacteria grow in filaments or masses of cells, but each cell in the colony is identical and capable of independent existence. The cells may be adjacent to one another because they did not separate after cell division or because they remained enclosed in a common sheath or slime secreted by the cells, but typically there is no continuity or communication between the cells.

PROCARYOTIC LIFE

The Bacteria are a group of single-cell microorganisms with procaryotic cellular configuration. The genetic material (DNA) of procaryotic cells exists unbound in the cytoplasm of the cells. There is no nuclear membrane, which is the definitive characteristic of eucaryotic cells such as those that make up, fungi, protista, plants and animals. Until recently, bacteria were the only known type of procaryotic cell, and the discipline of biology related to their study is called bacteriology. In the 1980's, with the outbreak of molecular techniques applied to phylogeny of life, another group of procaryotes was defined and informally named "archaebacteria".

This group of procaryotes has since been renamed Archaea and has been awarded biological Domain status on the level with Bacteria and Eucarya. The current science of bacteriology includes the study of both domains of procaryotic cells, but the name "bacteriology" is not likely to change to reflect the inclusion of archaea in the discipline. Actually, many archaea have been studied as intensively and as long as their bacterial counterparts, except with the notion that they were bacteria.

THE UNIVERSAL TREE OF LIFE

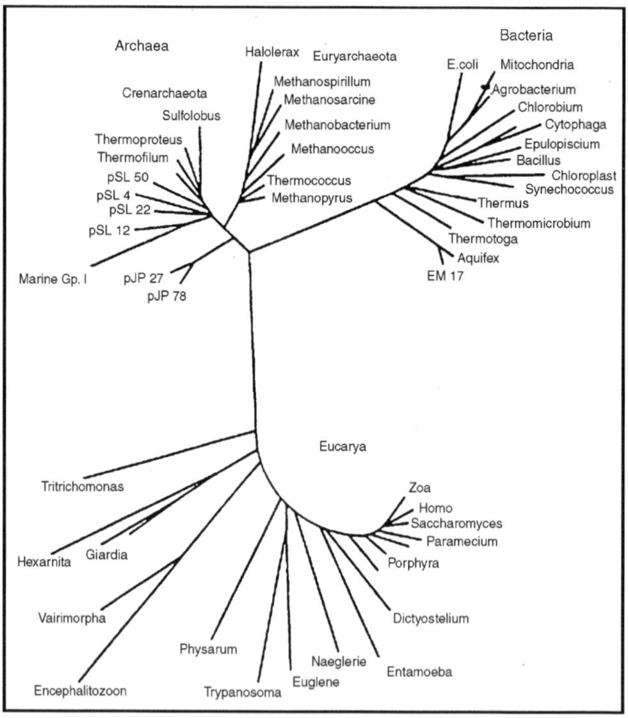

Fig. The Universal Tree of Life as Derived from Sequencing of ssrRNA. N. Pace. Note the Three Major Domains of Living Organisms: Archaea, Bacteria and Eucarya. The "Evolutionary Distance" between two Organisms is Proportional to the Measurable Distance between the end of a Branch to a Node to the end of a Comparative Branch. For Example, in Eucarya, Humans (*Homo*) are more Closely Related to Corn (*Zea*) than to Slime Molds *(Dictyostelium)*; or in Bacteria, *E. coli* is more Closely Related to *Agrobacterium* than to *Thermus.*

On the basis of small subunit ribosomal RNA (ssrRNA) analysis, the contemporary Tree of Life gives rise to three cellular "Domains": Archaea, Bacteria, and Eucarya. Bacteria (formerly known as eubacteria) and Archaea (formerly called archaebacteria) share the procaryotic type of cellular configuration, but otherwise are not related to one another any more closely than they are to the eucaryotic domain, Eucarya. Between the two procaryotes, Archaea are apparently more closely related to Eucarya than are the Bacteria. Eucarya consists of all eucaryotic cell-types, including protista, fungi, plants and animals.

BACTERIA

Current phylogenetic analysis of the Bacteria has demonstrated the existence of more than 20 distinct groups with taxonomic phylum status (Phylum is the highest taxon in a Domain).

Most groups listed consist of a distinct phylum; some are now subdivided into more than one phylum. Many groups consist of members that are phenotypically and physiologically unrelated. Furthermore, in some phylogenetic groups (*e.g.* Proteobacteria and Gram-positives) there is considerable phylogenetic diversity, unappreciated in the big Tree of Life.

Recognised Phyla of the Domain:

- Acidobacteria
- Actinobacteria
- Aquificae
- Bacteroidetes
- Chlamydiae
- Chlorobi
- Chloroflexi
- Chrysiogenetes
- Cyanobacteria
- Deferribacteres
- Deinococcus-Thermus
- Dictyoglomi
- Fibrobacteres
- Firmicutes
- Fusobacteria
- Gemmatimonadetes
- Nitrospirae
- Planctomycetes
- Proteobacteria
- Spirochaetes
- Thermodesulfobacteria
- Thermomicrobia

- Thermotogae
- Verrucomicrobia

Since we discuss the major groups of bacteria herein based on morphology, physiology, or ecology, it may be interesting to examine the phylogenetic relatedness of group members. Sometimes, members of a phylum are quite distinct (*e.g.*, spirochetes); other times, our organisation involves a confluence of phyla (*e.g.*, phototrophs, lithotrophs, pyogenic cocci). This makes for reflection on how related and unrelated organisms converge on the same strategy to adapt to specific environments.

Photosynthetic purple and green bacteria These bacteria conduct anoxygenic photosynthesis, also called bacterial photosynthesis. Bacterial photosynthesis differs from plant-type (oxygenic) photosynthesis in several ways. Bacterial photosynthesis does not produce O_2; in fact, it only occurs under anaerobic conditions. Bacterial photosynthesis utilises a type of chlorophyll other than chlorophyll *a,* and only one photosystem, photosystem I. The electron donor for bacterial photosynthesis is never H_2O but may be H_2, H_2S or S^o, or certain organic compounds. The light-absorbing pigments of the purple and green bacteria consist of bacterial chlorophylls and carotenoids. Phycobilins, characteristic of the cyanobacteria, are not found. Many purple and green sulphur bacteria store elemental sulphur as a reserve material that can be further oxidised to SO_4 as a photosynthetic electron donor.

The purple and green bacteria may use H_2S during photosynthesis in the same manner that cyanobacteria or algae or plants use H_2O as an electron donor for autotrophic CO_2 fixation (the "dark reaction" of photosynthesis). Or they may utilise organic compounds as electron donors for photosynthesis. For example, *Rhodobacter* can use light as an energy source while oxidising succinate or butyrate in order to obtain electrons for CO_2 fixation.

The bacterium that became an endosymbiont of eucaryotes and evolved into mitochondria is thought to be a relative of the purple non-sulfur bacteria. This conclusion is based on similar metabolic features of mitochondria and purple non-sulfur bacteria and on comparisons of the base sequences in their 16S rRNAs.

A. Purple sulphur bacteria (L to R): *Chromatium vinosum, Thiospirillum jenense, Thiopedia rosea.* The purple sulphur bacteria are classified among the Gamma Proteobacteria, a class that also includes *Pseudomonas* and *E. coli.*

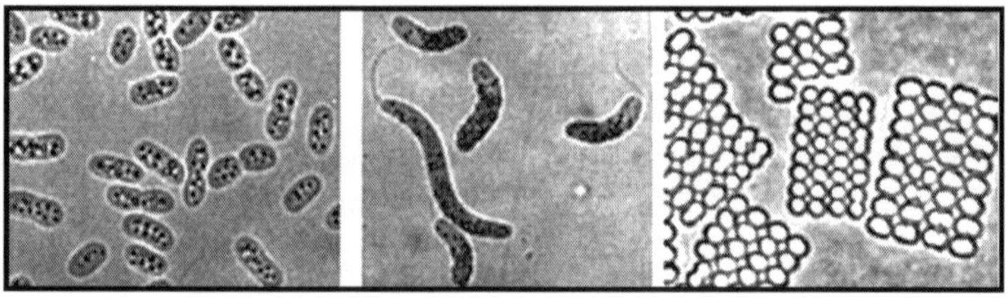

B. Purple non-sulfur bacteria (L to R): *Rhodospirillum rubrum, Rhodobacter sphaeroides, Rhodomicrobium vannielii.* The purple non-sulfur bacteria are in the Alpha Proteobacteria, which also includes *Rhizobium, Agrobacterium* and the Rickettsias. The latter bacteria represent a direct lineage to mitochondria.

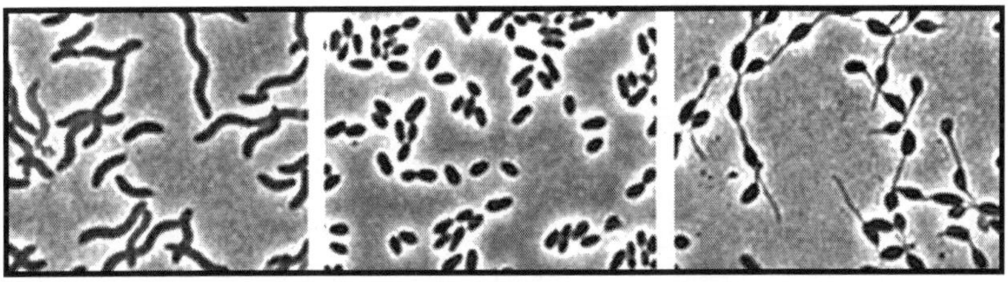

C. Green sulphur bacteria (L to R): *Chlorobium limicola, Prosthecochloris aestuarii, Pelodictyon clathratiforme.* The Green sulphur bacteria represent a distinct phylogenetic lineage and cluster in their own phylum represented by *Chlorobium* in Figure.

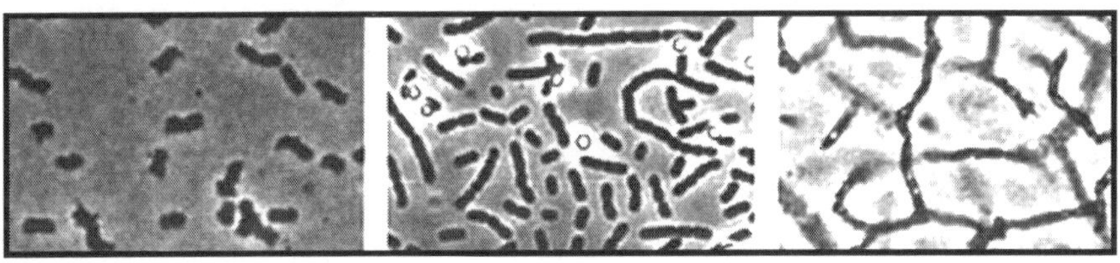

Fig. Photomicrographs (Phase Contrast and and Ordinary Illumination) of Various Photosynthetic Bacteria (Norbert Pfennig). Magnifications are about 1400X. The Purple and Green Bacteria Exhibit a Full Range of Procaryotic Morphologies, as these Photomicrographs Illustrate. Diversity among their Phylogenetic Relationships is also Noted.

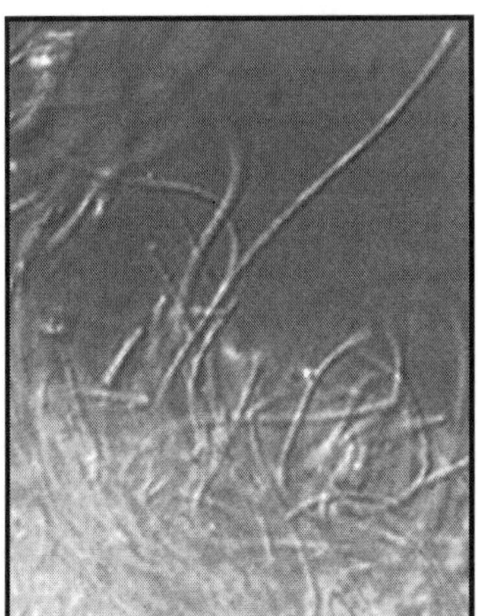

Fig. Green Non-sulfur Bacterium, *Chloroflexus* (T.D. Brock). *Chloroflexus* also Represents a Phylogenetically Distinct Group of Green Bacteria. *Chloroflexus* is a Thermophilic, Filamentous Gliding Bacterium.

Fig. Photosynthetic Procaryotes Growing in a Hot Spring Run-off Channel (T.D. Brock). The White Area of the Channel is Too Hot for Photosynthetic Life, but as the Water Cools Along a Gradient, the Coloured Phototrophic Bacteria Colonise and Ultimately Construct the Coloured Microbial Mats Composed of a Consortium of Photosynthetic Microorganisms.

Heliobacteria The heliobacteria are a unique group of anoxygenic photosynthetic bacteria. They have a typical bacterial PSI type reaction centre, but the primary pigment involved is bacteriochlorophyll g, which is unique to the group and has a unique absorption spectrum. This gives the heliobacteria their own environmental niche. Photosynthesis takes place at the cell membrane, which does not form folds or compartments as it does in purple phototrophic bacteria. Interestingly, phylogenetic analysis place the heliobacteria among the Firmicutes (Gram-positive bacteria), but they do not stain Gram-positive. They have no outer membrane and like certain other firmicutes (clostridia) they form heat resistant endospores. Also, are the only firmicutes known to conduct photosynthesis.

Heliobacteria are photoheterotrophic, requiring organic carbon sources, and they are exclusively anaerobic, heliobacteria have been found in soils and are apparently widespread in the waterlogged soils of paddy fields. They fix nitrogen and are probably important in the fertility of paddy fields.

Chloracidobacterium thermophilum Prospecting for genetic data in the microbial mats of Yellowstone Park and looking for evidence of a green sulphur bacteria, microbiologist David Ward of Montana State University found a "wildly different kind of phototrophic bacterium" in the 66°C water. Evidence for the organism was found using the techniques of metagenomics, and subsequently by generating an enriched monoculture. The organism described and grown was named *Chloracidobacterium thermophilum*. By demonstrating that the monoculture was tolerant to O_2, but could not be maintained in darkness, it was concluded that the organism was a new type of photosynthetic microbe.

Despite its affiliation with the bacterial phylum Acidobacterium, which consists of hyperthermophilic acidophiles, *Chloracidobacterium thermophilum* thrives in alkaline environments.

In addition to being able perform photosynthesis in the presence of oxygen, *Chloroacidobacterium thermophilum* is also the only oxygen-tolerant bacterium that constructs chlorosomes, providing light-harversting molecular antennae that allow the organisms to live at vanishingly low light intensities. Spirochetes are a phylogenetically distinct group of Bacteria which have a unique cell morphology and mode of motility. Spirochetes are very thin, flexible, spiral-shaped procaryotes that move by means of structures called axial filaments or endoflagella. The flagellar filaments are contained within a sheath between the cell wall peptidoglycan and an outer membrane. The filaments flex or rotate within their sheath which causes the cells to bend, flex and rotate during movement. Most spirochetes are free living (in muds and sediments), or live in associations with animals (*e.g.*, in the oral cavity or GI tract). A few are pathogens of animals (*e.g.*, leptospirosis in dogs, syphilis in humans and Lyme Disease in dogs and humans).

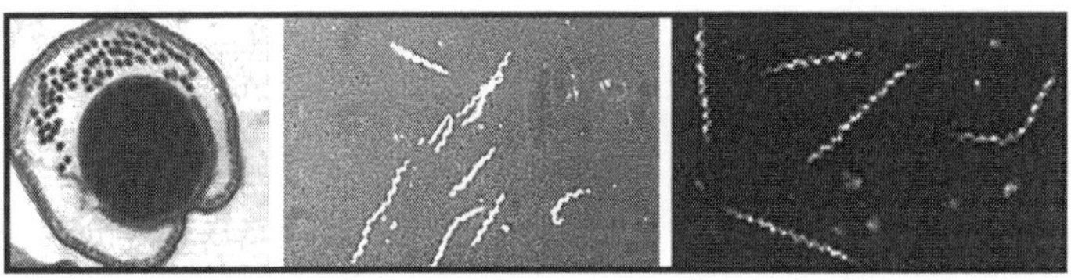

Fig. *Spirochetes:* A. Cross Section of a Spirochete Showing the Location of Endoflagella between the Inner Membrane and Outer Sheath; B. *Borrelia Burgdorferi,* the Agent of Lyme Disease; C. *Treponema Pallidum,* the Spirochete that Causes Syphilis.

Other spiral shaped and Curved Bacteria The main thing that unifies this group of bacteria is their spiral or vibrioid (curved) shape, although they are all classified among the Proteobacteria. Nonetheless, in certain environments, their characteristic shape can instantly inform an observer of their identity. Bacteria referred to as "spirilla" are Gram-negative aerobic heterotrophic bacteria with a helical or spiral shape.

Their metabolism is usually respiratory and never fermentative. Unlike spirochetes, they have a rigid cell wall and are motile by means of ordinary polar flagella. Spirilla are inhabitants of microaerophilic aquatic environments. Most spirilla require or prefer that oxygen in their environment be present in an amount that is well below atmospheric concentration.

The *Rhodospirillaceae are* found in the Alpha group of Proteobacteria; *Spirillaceae* and *Oceanospirillaceae* are Gamma Proteobacteria. As inhabitants of marine and fresh waters many spirilla are endowed with some interesting properties. *Magnetospirillum contains* magnetosomes and exhibits the property of magnetotaxis (movement in relationship to the magnetic field of the earth). *Oceanospirillum* lives in marine habitats and is able to grow at NaCl concentrations as high as 9 percent. *Azospirillum* is a nitrogen-fixing bacterium that enters into a mutualistic symbiosis with certain tropical grasses and grain crops. Spirilla are thought to play a significant role in recycling of organic matter, particularly in aquatic environments.

Two pathogens of humans are found among the spiral forms in the Epsilon group of Proteobacteria. *Campylobacter jejuni* is an important cause of bacterial diarrhea, especially in children. The bacterium is transmitted via contaminated food, usually undercooked poultry or shellfish, or untreated drinking water. *Helicobacter pylori* is able to colonise the gastric mucosal cells of humans, *i.e.*, the lining of the stomach, and it has been well established as the cause of peptic ulcers. Bacteria with a curved rod or comma shape are referred to as "vibrios". Like the spiral forms, vibrios are very common bacteria in aquatic environments. They are found among the Gamma Proteobacteria and have structural and metabolic properties that overlap with both the enterics and the pseudomonads. In Bergey's Manual (2001) *Vibrionaceae* is a family on the level with *Enterobacteriaceae*.

Vibrios are facultative like enterics, but they have polar flagella, are oxidase-positive, and dissimilate sugars in the same manner as the pseudomonads. In aquatic habitats they overlap with the *Pseudomonadaceae* in their ecology, although *Pseudomonas* species favour fresh water and vibrios prefer salt water. The genus *Vibrio* contains an important pathogen of humans, *Vibrio cholerae,* the cause of Asiatic cholera. Cholera is an intestinal disease with a pathology related to diarrheal diseases caused by the enteric bacteria.

Five species of marine vibrios exhibit the property of bioluminescence, the ability to emit light of a blue-green colour. These bacteria may be found as saprophytes of dead fish or as symbionts of living fish and invertebrates in marine environments. Some grow in special organs of the fish and emit light for the benefit of the fish (to attract prey, or as a mating signal) in return for a protected habitat and supply of nutrients. The reaction leading to light emission, catalysed by the enzyme luciferase, has been found to be the same in all procaryotes, and differs from light emission by eucaryotes such as the fire fly. Luciferase diverts electrons from the normal respiratory electron transport chain and causes formation of an excited peroxide that leads to emission of light. The small vibrioid bacterium, *Bdellovibrio, is* a tiny curved rod that is a parasite of other Gram-negative bacteria, including *E. coli*. It preys on other bacteria by entering into the periplasmic space and obtaining nutrients from the cytoplasm of its host cell while undergoing an odd type of reproductive cycle. *Bdellovibrio*is a member of the Delta Proteobacteria.

Myxobacteria are a group of fruiting gliding bacteria that comprise a unique order of Delta Proteobacteria. They exhibit a unique type of gliding motility. The vegetative cells move (glide) about together as a swarm, and then they aggregate together to form a multicellular fruiting body in which development and spore formation takes place. They exhibit the most complex behavioural patterns and life cycles of all known procaryotes. Myxobacteria are inhabitants of the soil. They have a eucaryotic counterpart in nature in the *Myxomycetes*, or slime molds, and the two types of organisms are an example of parallel or convergent evolution, having adopted similar life styles in the soil environment.

The vegetative cells of myxobacteria are typical Gram-negative rods that glide across a substrate such as a decaying leaf or piece of animal dung, or colonies of other bacteria. They obtain nutrients from the substrate as they glide across it and they secrete a slime track which other myxobacterial cells preferentially follow. If their nutrients become exhausted, the cells signal to one another to aggregate and form a swarm of myxobacteria which eventually differentiate into a multicellular fruiting body that contains myxospores, a type of dormant cell descended from a differentiated vegetative cell. In the case of *Stigmatella*, the myxospores are packed into secondary structures called cysts, which develop at the tips of the fruiting body. The bright-coloured fruiting bodies of myxobacteria, containing millions of cells and spores, can often be seen with the unaided eye on dung pellets and decaying vegetation in the soil.

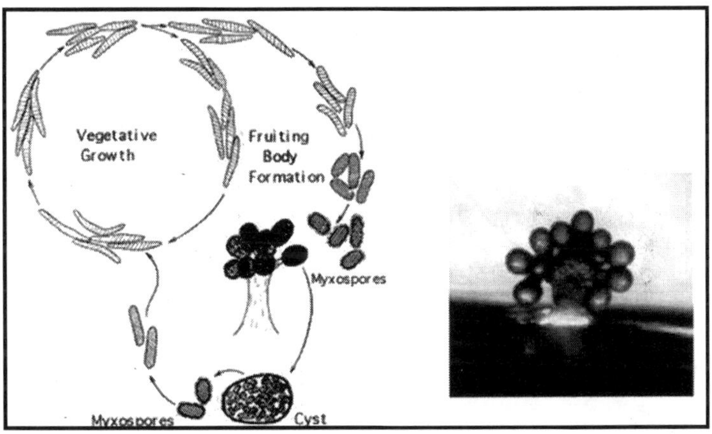

Fig. *Stigmatella Aurantiaca,* a Fruiting Myxobacterium: L. Life Cycle R. Fruiting Body.

Lithotrophs Lithotrophy, a type of metabolism that requires inorganic compounds as sources of energy. This metabolism is firmly established in both the Archaea and the Bacteria. The methanogens utilise H_2 as an energy source, and many extreme thermophiles use H_2S or elemental sulphur as a source of energy for growth. Lithotrophic Bacteria are typically Gram-negative species that utilise inorganic substrates including H_2, NH_3, NO_2, H_2S, S, Fe^{++}, and CO. Ecologically, the most important lithotrophic Bacteria are the nitrifying bacteria, *Nitrosomonas* and *Nitrobacter* that together convert NH_3 to NO_2, and NO_2 to NO_3, and the colourless sulphur bacteria, such as *Thiobacillus,* that oxidise H_2S to S and S to SO_4. Most lithotrophic bacteria are autotrophs, and in some cases, they may play an important role in primary production of organic material in nature.

Lithotrophic metabolism does not extend to eucaryotes (unless a nucleated cell harbours lithotrophic endosymbiotic bacteria), and these bacteria are important in the biogeochemical cycles of the elements.

Fig. Lithotroph Habitats. A. Stream in Northern Wisconsin Near Hayward is a Good Source of Iron Bacteria (John Lindquist). B. Bacteriologist J.C. Ensign of the University of Wisconsin Observing Growth of Iron Bacteria in a Run-off Channel from the Chocolate Pots along the Gibbon River, in Yellowstone National Park (K.Todar). C. An Acid Hot Spring at the Norris Geyser Basin in Yellowstone is Rich in Iron and Sulphur (T.D. Brock). D. A Black Smoker Chimney in the Deep Sea Emits Iron Sulfides at very High Temperatures (270 to 380 degrees C).

Pseudomonads. "Pseudomonad" is an informal term for bacteria which morphologically and physiologically resemble members of the genus *Pseudomonas.* They are unified only as Gram-negative rods with a strictly-respiratory mode of metabolism. The term has been applied to membersof the genera *Pseudomonas,* and *Xanthomonas,* which are Alpha Proteobacteria, and to plant and animal pathogens such as *Burkholderia, Ralstonia* and *Acidovorax,* which are Beta Proteobacteria although other related and non-related bacteria share their definitive characteristics, *i.e.*, Gram-negative aerobic rods.

The morphology and habitat of many pseudomonads sufficiently overlaps with the enterics (below) so that microbiologists must quickly learn how to differentiate these two types of Gram-negative motile rods. Pseudomonads move by polar flagella; enterics such as *E. coli* swim by means of peritrichous flagella. Enterics ferment sugars such as glucose; pseudomonads generally do not ferment sugars. And most pseudomonads have an unusual cytochrome in their respiratory electron transport chain that can be detected in colonies by a colourimetric test called the oxidase test. Pseudomonads are typically oxidase- positive.

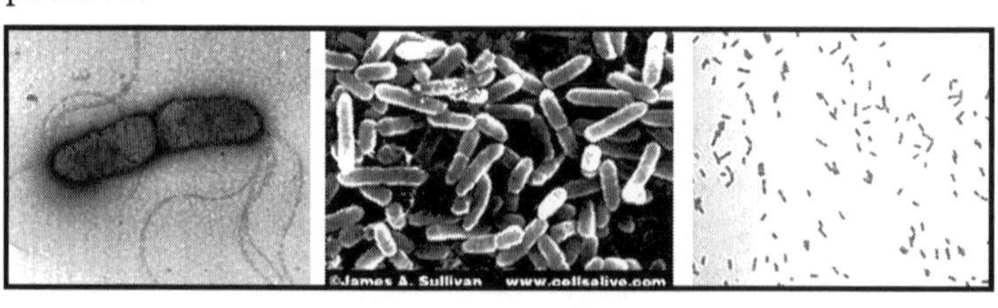

Fig. Profile of a Pseudomonad: Gram-negative Rods Motile by Polar Flagella. A. Electron Micrograph, Negative Stain. B. Scanning Electron Micrograph. C. Gram stain.

Most pseudomonads are free-living organisms in soil and water; they play an important role in decomposition, biodegradation, and the C and N cycles. The phrase "no naturally-occurring organic compound cannot be degraded by some microorganism" must have been coined to apply to members of the genus *Pseudomonas,* known for their ability to degrade hundreds of different organic compounds including insecticides, pesticides, herbicides, plastics, petroleum substances, hydrocarbons and other of the most refractory molecules in nature. However, they are usually unable to degrade biopolymers in their environment, such as cellulose and lignin, and their role in anaerobic decomposition is minimal.

There are about 150 species of *Pseudomonas*, but, especially among the plant pathogens, there are many strains and biovars among the species. These bacteria are frequently found as part of the normal flora of plants, but they are one of the most important bacterial pathogens of plants, as well. *Pseudomonas syringae* and *Xanthomonas* species cause a wide variety of plant diseases as discussed below. One strain of *Pseudomonas* that lives on the surfaces of plants can act as an "ice nucleus" which causes ice formation and inflicts frost damage on plants at one or two degrees *above* the conventional freezing temperature of water (0 degrees C). One *Pseudomonas* species is an important pathogen of humans, *Pseudomonas aeruginosa,* the quintessential opportunistic pathogen, which is a leading cause of hospital-acquired (nosocomial) infections.

Among some interesting or important ecologic relatives of the pseudomonads are *Rhizobium and Bradyrhizobium,* species that fix nitrogen in association with leguminous plants, and related *Agrobacterium* species that cause tumours ("galls") in plants. These bacteria are discussed later in this article because of their special relationships with plants. Relatives of the pseudomonads also include the methanotrophs that can oxidise methane and other one-carbon compounds, the azotobacters, which are very prevalent free-living (non-symbiotic) nitrogen-fixing bacteria.

Enterics. The Enteric Bacteria are Gram-negative rods with facultative anaerobic metabolism that live in the intestinal tracts of animals. This group consists of *Escherichia coli* and its relatives, the members of the family *Enterobacteriaceae.* Enteric bacteria are related phenotypically to several other genera of bacteria such as *Pseudomonas* and *Alcaligenes,* but are physiologically quite unrelated. Generally, a distinction can be made on the ability to ferment glucose: enteric bacteria all ferment glucose to acid end products while similar Gram-negative bacteria cannot ferment glucose. Because they are consistent members of the normal flora of humans, and because of their medical importance, an extremely large number of enteric bacteria have been isolated and characterised.

Escherichia coli is, of course, the type species of the enterics. *E. coli* is such a regular inhabitant of the intestine of humans that it is used by public health authorities as an indicator of fecal pollution of drinking water supplies, swimming beaches, foods, etc. *E. coli* is the most studied of all organisms in biology because of its occurrence, and the ease and speed of growing the bacteria in the laboratory.

It has been used in hundreds of thousands of experiments in cell biology, physiology, and genetics, and was among the first cells for which the entire chromosomal DNA base sequence was determined. In spite of the knowledge gained about the molecular biology and physiology of *E. coli,* surprisingly little is known about its ecology, for example why it consistently associates with humans, how it helps its host, how it harms its host, etc. A few strains of *E. coli* are pathogenic (one is notorious, strain 0157:H7, that keeps turning up in raw hamburger headed for a fast-food restaurants). Pathogenic strains of *E. coli* cause intestinal tract infections (usually acute and uncomplicated, except in the very young.

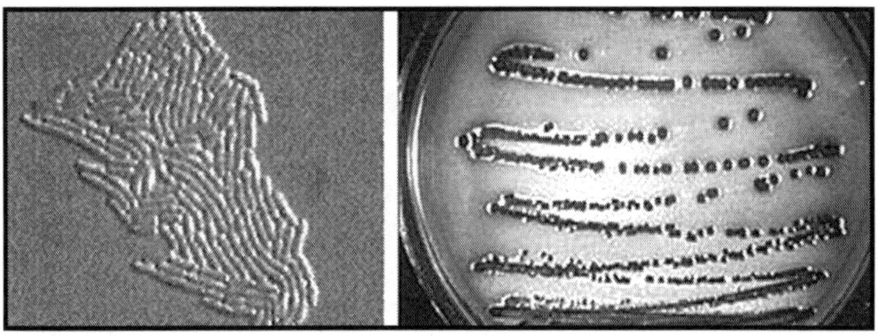

Fig. Left: *Escherichia Coli* cells. Right: *E. Coli* colonies on EMB Agar.

The enteric group also includes some other intestinal pathogens of humans such as *Shigella dysenteriae*, cause of bacillary dysentery (*Shigella* and Shigellosis), and *Salmonella enterica*, cause of gastroenteritis. *Salmonella typhi*, which infects via the intestinal route, causes typhoid fever (*Salmonella* and Salmonellosis).

Some bacteria that don't have an intestinal habitat resemble *E. coli* in enough ways to warrant inclusion in the enteric group. This includes *Proteus,* a common saprophyte of decaying organic matter, *Yersinia pestis,* which causes bubonic plague, and *Erwinia,* an important pathogen of plants.

Gram-negative pathogens. Most, but not all, of the Gram negative bacteria that are important pathogens of humans are found scattered throughout the Proteobacteria. In the Alpha Proteobacteria, one finds the Rickettsias, a group of obligate intracellular parasites which are the cause of typhus and Rocky Mountain Spotted Fever. In the Beta group, the agents of whooping cough (pertussis) (*Bordetella pertussis),* (*Neisseria gonorrhoeae*), and meningococcal meningitis (*Neisseria meningitidis*) are found. (Gonorrhea and Meningitis) Among the Gamma group, *Pseudomonas aeruginosa,* the enterics, and *Vibrio cholerae* have already been mentioned. Likewise, the agents of Legionnaire's' pneumonia (*Legionella pneumophilia),* and childhood meningitis *(Haemophilus influenzae)* are Gamma Proteobacteria. *Campylobacter* and *Helicobacter* are

Epsilon Proteobacteria. Most of these bacteria are discussed elsewhere in this article and/or in separate chapters which deal with their pathogenicity for humans.

Nitrogen-fixing organisms. This is a diverse group of procaryotes, reaching into phylogenetically distinct groups of Archaea and Bacteria. Members are unified only on the basis of their metabolic ability to "fix" nitrogen. Nitrogen fixation is the reduction of N_2 (atmospheric nitrogen) to NH_3 (ammonia). It is a complicated enzymatic process mediated by the enzyme nitrogenase. Nitrogenase is found only in procaryotes and is second only to RUBP carboxylase (the enzyme responsible for CO_2 fixation) as the most abundant enzyme on Earth.

The conversion of nitrogen gas (which constitutes about 80 percent of the atmosphere) to ammonia introduces nitrogen into the biological nitrogen cycle. Living cells obtain their nitrogen in many forms, but usually from ammonia (NH_3) or nitrates (NO_3), and never from N_2. Nitrogenase extracts N_2 from the atmosphere and reduces it to NH_3 in a reaction that requires substantial reducing power (electrons) and energy (ATP). The NH_3 is immediately assimilated into amino acids and proteins by subsequent cellular reactions. Thus, nitrogen from the atmosphere is fixed into living (organic) material.

Although a widespread trait in procaryotes, nitrogen fixation occurs in only a few select genera. Outstanding among them are the symbiotic bacteria *Rhizobium* and *Bradyrhizobium* which form nodules on the roots of legumes. In this symbiosis the bacterium invades the root of the plant and fixes nitrogen which it shares with the plant. The plant provides a favourable habitat for the bacterium and supplies it with nutrients and energy for efficient nitrogen fixation. *Rhizobium* and *Bradyrhizobium* are Gram-negative aerobes related to the pseudomonads (above). An unrelated bacterium, an actinomycete (below), enters into a similar type of symbiosis with plants. The actinomycete, *Frankia,* forms nodules on the roots of several types of trees and shrubs, including alders (*Alnus*), wax myrtles (*Myrica*) and mountain lilacs (*Ceanothus*). They, too, fix nitrogen which is provided to their host in a useful form. This fact allows alder species to be "pioneer plants" (among the first to colonise) in newly-forming nitrogen-deficient soils. Still other bacteria live in regular symbiotic associations with plants on roots or leaves and fix nitrogen for their hosts, but they do not cause tissue hyperplasia or the formation of nodules.

Cyanobacteria are likewise very important in nitrogen fixation. Cyanobacteria provide fixed nitrogen, in addition to fixed carbon, for their symbiotic partners which make up lichens. This enhances the capacity for lichens to colonise bare areas where fixed nitrogen is in short supply. In some parts of Asia, rice can be grown in the same paddies continuously without the addition of fertilizers because of the presence of nitrogen fixing cyanobacteria. The cyanobacteria, especially *Anabaena*, occur in association with the small floating water fern *Azolla,* which forms masses on the paddies. Because of the nearly obligate association of *Azolla* with *Anabaena,* paddies covered with *Azolla* remain rich in fixed nitrogen.

In addition to symbiotic nitrogen-fixing bacteria, there are various free-living nitrogen-fixing procaryotes in both soil and aquatic habitats. Cyanobacteria may be able to fix nitrogen in virtually all habitats that they occupy. Clostridia and some methanogens fix nitrogen in anaerobic soils and sediments, including thermophilic environments.

A common soil bacterium, *Azotobacter* is a vigourous nitrogen fixer, as is *Rhodospirillum*, a purple sulphur bacterium. Even *Klebsiella*, an enteric bacterium closely related to *E. coli,* fixes nitrogen. There is great scientific interest, of course, in knowing how one might move the genes for nitrogen fixation from a procaryote into a eucaryote such as corn or some other crop plant. The genetically engineered plant might lose its growth requirement for costly ammonium or nitrate fertilizers and grow in nitrogen deficient soils.

Besides nitrogen fixation, bacteria play other essential roles in the processes of the nitrogen cycle. For example, saprophytic bacteria, decompose proteins releasing NH_3 in the process of ammoniafication. NH_3 is oxidised by lithotrophic *Nitrosomonas* species to NO_2 which is subsequently oxidised by *Nitrobacter* to NO_3. The overall conversion of NH_3 to NO_3 is called nitrification. NO_3 can be assimilated by cells as a source of nitrogen (assimilatory nitrate reduction), or certain bacteria can reduce NO_3 during a process called anaerobic respiration, wherein nitrate is used in place of oxygen as a terminal electron acceptor for a process analogous to aerobic respiration. In the case of anaerobic respiration, NO_3 is first reduced to NO_2, which is subsequently reduced to N_2O or N_2 (all gases).

This process is called denitrification and it occurs in anaerobic environments where nitrates are present. If denitrification occurs in crop soils it may not be beneficial to agriculture if it converts utilisable forms of nitrogen (as in nitrate fertilizers) to nitrogen gases that will be lost into the atmosphere. A related process call dissimilatory nitrate reduction, conducted by certain *Bacillus* species, reduces NO_3 to NH_3. One rationale for tilling the soil is to keep it aerobic, in order to discourage these facultative processes in *Pseudomonas* and *Bacillus*, which are ubiquitous inhabitants.

The pyogenic cocci are spherical bacteria which cause various suppurative (pus-producing) infections in animals. Included are the Gram-positive cocci *Staphylococcus aureus, Streptococcus pyogenes* and *Streptococcus pneumoniae,* and the Gram-negative cocci, *Neisseria gonorrhoeae* and *N. meningitidis.* These bacteria are leading pathogens of humans. It is estimated that they produce at least a third of all the bacterial infections of humans, including strep throat, pneumonia, food poisoning, various skin diseases and severe types of septic shock, gonorrhea and meningitis. *Staphylococcus aureus* is arguably the most successful of all bacterial pathogens because it has a very wide range of virulence determinants (so it can produce a wide range of infections) and it often occurs as normal flora of humans (on skin, nasal membranes and the GI tract), which ensures that it is readily transmitted from one individual to another. In terms of their phylogeny, physiology and genetics, these genera of bacteria are quite unrelated to one another. They share a common ecology, however, as parasites of humans.

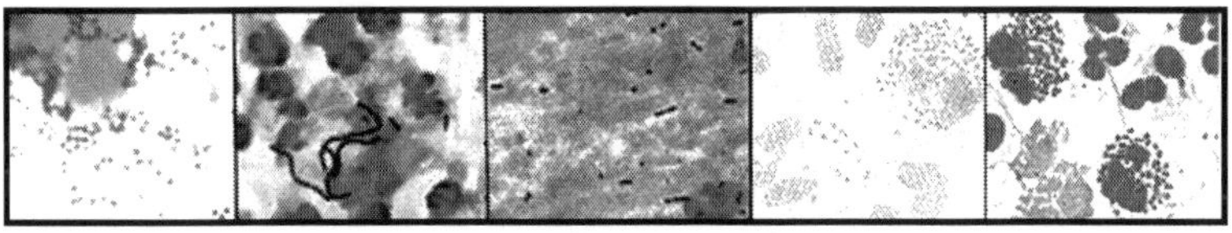

Fig. Gallery of Pyogenic Cocci, Gram Stains of Clinical Specimens (pus), L to R: *Staphylococcus Aureus, Streptococcus Pyogenes, Streptococcus Pneumoniae, Neisseria Gonorrhoeae, Neisseria Meningitidis*. The Large Cells with Lobed Nuclei are Neutrophils. Pus is the Outcome of the Battle between Phagocytes (Neutrophils) and the Invading Cocci. As the Bacteria are Ingested and Killed by the Neutrophils, the Neutrophils Eventually Lyse (Rupture) and Release their Own Components, Plus the Digested Products of Bacterial Cells, which are the Make-up of Pus. As a Defence against Phagocytes the Staphylococci and Streptococci Produce Toxins that Kill the Neutrophils before they are Able to Ingest the Bacteria. This Contributes to the Pus, and therefore these Bacteria are "Pyogenic" During their Pathogenic Invasions.

Two species of Staphylococcus live in association with humans: *Staphylococcus epidermidis* which lives normally on the skin and mucous membranes, and *Staphylococcus aureus* which may occur normally at various locales, but in particular on the nasal membranes (nares). *S. epidermidis* is rarely a pathogen and probably benefits its host by producing acids on the skin that retard the growth of dermatophytic fungi. *Staphylococcus aureus* always has the potential to cause disease and so is considered a pathogen.

Different strains of *S. aureus* differ in the range of diseases they can cause, including boils and pimples, wound infections, pneumonia, osteomyelitis, septicemia, food intoxication, and toxic shock syndrome. *S. aureus* is the leading cause of nosocomial (hospital-acquired) infections by Gram-positive bacteria. Also, it is notoriously resistant to penicillin and many other antibiotics. Recently, a strain of *S. aureus* has been reported that is resistant to EVERY known antibiotic in clinical usage, which is a grim reminder that the clock is ticking on the lifetime of the usefulness of current antibiotics in treatment of infectious disease.

Streptococcus pypgenes, more specifically the Beta-hemolytic Group A Streptococci, like *S. aureus*, causes an array of suppurative diseases and toxinoses (diseases due to the production of a bacterial toxin), in addition to some autoimmune or allergic diseases. *S. pyogenes* is rarely found as normal flora (<1 per cent), but it is the main streptococcal pathogen for man, most often causing tonsillitis or strep throat. Streptococci also invade the skin to cause localised infections and lesions, and produce toxins that cause scarlet fever and toxic shock. Sometimes, as a result of an acute streptococcal infection, anomalous immune responses are started that lead to diseases like rheumatic fever and glomerulonephritis, which are called post-streptococcal sequelae. Unlike the staphylococci, the streptococci have not developed widespread resistance to penicillin and the other beta lactam antibiotics, so that the beta lactams remain drugs of choice for the treatment of acute streptococcal infections.

Streptococcus pneumoniae is the most frequent cause of bacterial lobar pneumonia in humans. It is also a frequent cause of otitis media (infection of the middle ear) and meningitis. The bacterium colonises the nasopharynx and from there gains access to the lung or to the eustachian tube. If the bacteria descend into the lung they can impede engulfment by alveolar macrophages if they possess a capsule which somehow prevents the engulfment process. Thus, encapsulated strains are able to invade the lung and are virulent (cause disease) and non-capsulated strains, which are readily removed by phagocytes, are non-virulent.

The *Neisseriaceae* comprise a family of Gram-negative Beta Proteobacteria with metabolic characteristics similar to pseudomonads. The neisseriae are small, Gram-negative cocci usually seen in pairs with flattened adjacent sides. Most neisseriae are normal flora or harmless commensals of mammals living on mucous membranes.

In humans they are common residents of the throat and upper respiratory tract. Two species are primary pathogens of humans, *Neisseria gonorrhoeae* and *Neisseria meningitidis,* the bacterial causes of gonorrhea and meningococcal meningitis.

Neisseria gonorrhoeae is the second leading cause of sexually-transmitted disease in the U.S., causing over three million cases of gonorrhea annually. Sometimes, in females, the disease may be unrecognised or asymptomatic such that an infected mother can give birth and unknowingly transmit the bacterium to the infant during its passage through the birth canal. The bacterium is able to colonise and infect the newborn eye resulting neo-natal ophthalmia, which may produce blindness. For this reason (as well as to control Chlamydia which may also be present), an antimicrobial agent is usually added to the neo-nate eye at the time of birth.

Neisseria meningitidis is one bacterial cause of meningitis, an inflammation of the meninges of the brain and spinal cord. Other bacteria that cause meningitis include *Haemophilus influenzae*, *Staphylococcus aureus* and *Escherichia coli*. Meningococcal meningitis differs from other causes in that it is often responsible for epidemics of meningitis. It occurs most often in children aged 6 to 11 months, but it also occurs in older children and in adults. Meningococcal meningitis can be a rapidly fatal disease, and untreated meningitis has a mortality rate near 50 percent. However, early intervention with antibiotics is highly effective, and with treatment most individuals recover without permanent damage to the nervous system. Lactic acid bacteria are Gram-positive, non-sporeforming rods and cocci which produce lactic acid as a sole or major end product of fermentation. They are important in the food industry as fermentation organisms in the production of cheese, yogurt, buttermilk, sour cream, pickles, sauerkraut, sausage and other foods. Important genera are *Streptococcus, Enterococcus, Lactococcus* and *Lactobacillus*.

Some species are normal flora of the human body (found in the oral cavity, GI tract and vagina); some streptococci are pathogens of humans. Certain oral lactic acid bacteria are responsible for the formation of dental plaque and the initiation of dental caries (cavities). *Enterococcus faecalis* is a consistent member of the GI tract of humans and an important purveyor of drug resistance via horizontal gene transmission in the intestine.

Lactococcus lactis has been nominated as the "state microbe" of Wisconsin. The bacterium is one of the most important microbes involved in the dairy industry. It is non-pathogenic and critical for manufacturing dairy products like buttermilk, yogurt and cheese. When *L. lactis* ssp. lactis is added to milk, the lactic acid produced by the bacterium curdles the milk that then separates to form curds, which are used to produce cheese and whey.

Lactococcus lactis is also used to prepare pickled vegetables, beer, wine, some breads and sausages and other fermented foods. Researchers anticipate that understanding the physiology and genetic make-up of this bacterium will prove invaluable for food manufacturers as well as the pharmaceutical industry, which is exploring the capacity of *L. lactis* to serve as a vehicle for delivering drugs.

Endospore-forming bacteria produce a unique resting cell called an endospore. They are Gram-positive and usually rod-shaped, but there are exceptions. The two important genera are *Bacillus*, the members of which are aerobic endospore-forming bacteria in the soils, and *Clostridium,* whose species are anaerobic sporeformers of soils, sediments and the intestinal tracts of animals.

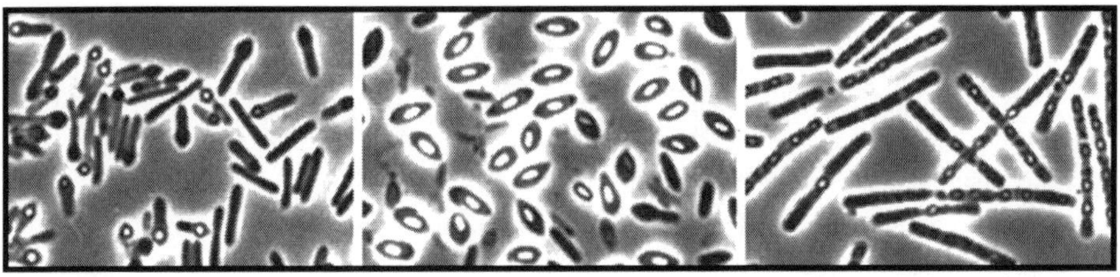

Fig. Endospore-forming Bacilli (Phase Contrast Illumination). Endospores are Dehydrated, Refractile Cells Appearing as Points of Bright Light Under Phase Microscopy. Endospore-forming Bacteria are Characterised by the Location (Position) of the Endospore in the Mother Cell (Sporangium) before its Release. The Spore may be Central, Terminal or Subterminal, and the Sporangium may or may not be Swollen to Accommodate the Spore.

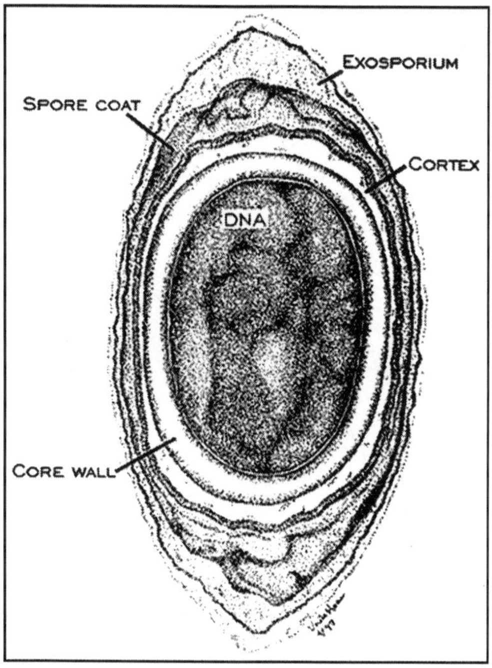

Fig. Anatomy of an Endospore, Cross Section Drawing by Viake Haas. Endospores Differ from the Vegetative Cells that Form them in a Variety of Ways. Several New Surface Layers Develop Outside the Core (Cell) Wall, Including the Cortex and Spore Coat. The Cytoplasm is Dehydrated and Contains only the Cell Genome and a Few Ribosomes and Enzymes. The Endospore is Cryptobiotic (Exhibits No Signs of Life) and is Remarkably Resistant to Environmental Stress such as Heat (Boiling), Acid, Irradiation, Chemicals and Disinfectants. Some Endospores have Remained Dormant for 25 Million Years Preserved in Amber, Only to be Shaken Back into Life when Extricated and Introduced into a Favourable Environment.

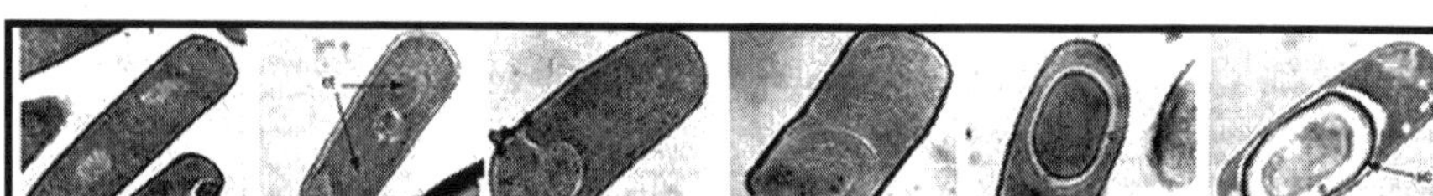

Fig. The Sequential Steps in the Process of Endospore Formation in *Bacillus Subtilis.*

Some sporeformers are pathogens of animals, usually due to the production of powerful toxins. *Bacillus anthracis* causes anthrax, a disease of domestic animals (cattle, sheep, etc.) which may be transmitted to humans.

Bacillus cereus is becoming increasingly recognised as an agent of food poisoning. *Clostridium botulinum* causes botulism a form of food-poisoning, and *Clostridium tetani* causes tetanus.

In association with the process of sporulation, some *Bacillus* species form a crystalline protein inclusion called parasporal crystals. The protein crystal and the spore (actually the spore coat) are toxic to lepidopteran insects (certain moths and caterpillars) if ingested. The crystals and spores of *Bacillus thuringiensis* are marketed as "Bt" a natural insecticide for use on garden or crop plants. Another species of *Bacillus*, *B. cereus*, produces an antibiotic that inhibits growth of *Phytophthera,* a fungus that attacks alfalfa seedling roots causing a "damping off" disease. The bacteria, growing in association with the roots of the seedlings, can protect the plant from disease.

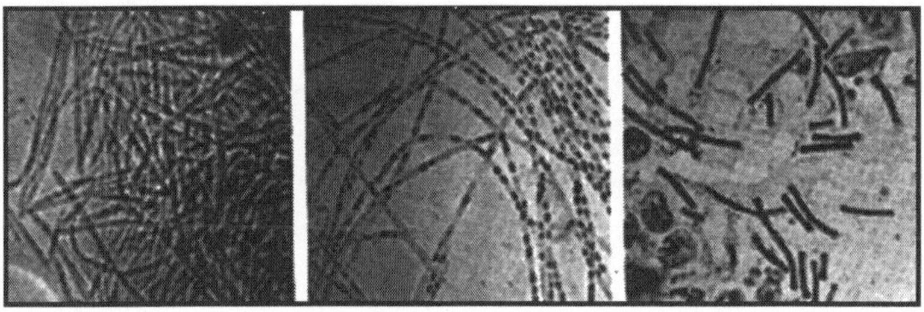

Fig. Robert Koch's Original Photomicrographs of *Bacillus Anthracis.* In 1876, Koch Established by Careful Microscopy that the Bacterium was Always Present in the Blood of Animals that Died of Anthrax. He took a Small Amount of Blood from such an Animal and Injected it into a Healthy Mouse, which Subsequently became Diseased and Died. He took Blood from that Mouse and Injected it into a another Healthy Mouse. After Repeating this Several Times he was Able to Recover the Original Anthrax Organism from the Dead Mouse, Demonstrating for the First Time that a Specific Bacterium is the Cause of a Specific Disease. In so doing, he Established Koch's Postulates, which Still Today Supply the Microbiological Standard to Demonstrate that a Specific Microbe is the Cause of a Specific Disease.

Also, apparently in association with the sporulation process, some bacilli produce clinically-useful antibiotics. *Bacillus* and *Paenibacillus* antibiotics, such as bacitracin and polymyxin are usually polypeptide molecules that contain unusual amino acids.

Actinomycetes and related bacteria are a large group of Gram-positive bacteria that usually grow by filament formation, or at least show a tendency towards branching and filament formation.

Many of the organisms can form resting structures called spores, but they are not the same as endospores. Branched forms superficially resemble molds and are a striking example of convergent evolution of a procaryote and a eucaryote together in the soil habitat.

Actinomycetes such as *Streptomyces* have a world-wide distribution in soils. They are important in aerobic decomposition of organic compounds and have an important role in biodegradation and the carbon cycle. Products of their metabolism, called geosmins, impart a characteristic earthy odour to soils.

Actinomycetes are the main producers of antibiotics in industrial settings, being the source of most tetracyclines, macrolides (*e.g.*, erythromycin), and aminoglycosides (*e.g.*, streptomycin, gentamicin, etc.). Two bacteria in this diverse group are important pathogens of humans: *Mycobacterium tuberculosis* is the cause of tuberculosis; *Corynebacterium diphtheriae* is the cause of diphtheria. Also, many non-pathogenic mycobacteria and corynebacteria live in associations with animals.

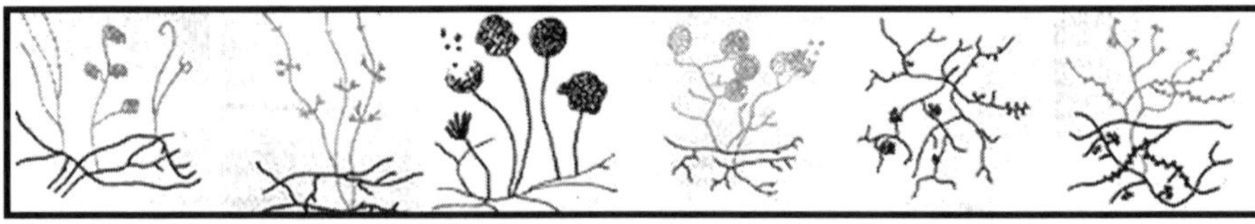

Fig. Schematic Diagrams Illustrating Mycelial Growth and Spore Formation in Several Genera of Actinomycetes.

Rickettsias and chlamydiae are two unrelated groups of Bacteria that are obligate intracellular parasites of eucaryotic cells. Rickettsias cannot grow outside of a host cell because they have leaky membranes and are unable to obtain nutrients in an extracellular habitat. Chlamydiae are unable to produce ATP in amounts required to sustain metabolism outside of a host cell and are, in a sense, energy-parasites.

Rickettsias occur in nature in the gut lining of arthropods (ticks, fleas, lice, etc.). They are transmitted to vertebrates by an arthropod bite and produce such diseases as typhus fever, Rocky Mountain Spotted Fever, Q fever and canine ehrlichiosis. Chlamydiae are tiny bacteria that infect birds and mammals. They may colonise and infect tissues of the eye and urogenital tract in humans. *Chlamydia trachomatis* causes several important diseases in humans: chlamydia, the most prevalent sexually transmitted disease in the U.S., trachoma, a leading cause of blindness worldwide, and lymphogranuloma venereum.

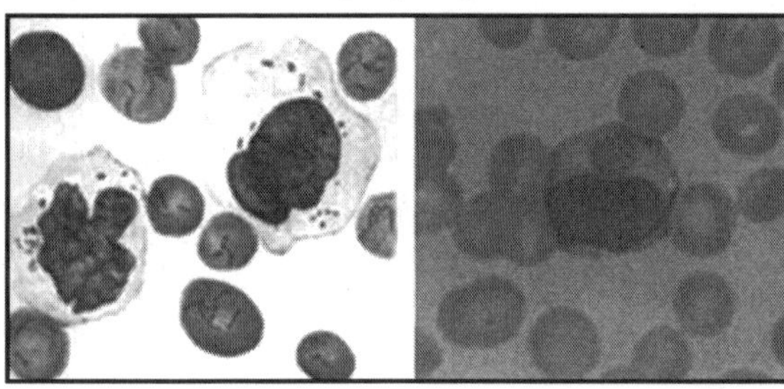

Fig. Mammalian Cells Infected with Rickettsial Organisms. L. *Bartonella Bacilliformis* Infection of Human Erythrocytes and Blood Monoctyes. R. *Ehrlichia Canis* Infection of Canine Erythrocytes and Blood Monocytes. The Distinct Stained Intracytoplasmic Inclusion Body in the Monocyte is Characteristic of the Infection.

Mycoplasmas are a group of bacteria that lack a cell wall. The cells are bounded by a single triple-layered membrane. They may be free-living in soil and sewage, parasitic inhabitants of the mouth and urinary tract of humans, or pathogens in animals and plants. In humans, *Mycoplasma pneumoniae* causes primary atypical pneumonia ("walking pneumonia").

Mycoplasmas include the smallest known cells, usually about 0.2 - 0.3 micrometers in diameter. Mycoplasmas correspondingly have the smallest known genome of any cell. Their DNA is thought to contain about 650 genes, which is about one-fifth the number found in *E. coli* and other common bacteria. Mycoplasmas can survive without a cell wall because their cytoplasmic membrane is more stable than that of other procaryotes. In one group of mycoplasmas, the membrane contains sterols which seem to be responsible for the stability. Also, mycoplasmas tend to inhabit environments of high osmolarity wherein the risk of osmotic shock and lysis of the cells is minimized. Plant-pathogenic bacteria. Many economically-important diseases of plants are caused by members of the Bacteria. It is estimated that one-eighth of the crops worldwide are lost to diseases caused by bacteria, fungi or insects. Almost all kinds of plants can be affected by bacterial diseases, and many of these diseases can be extremely destructive.

Almost all plant-pathogenic bacteria are Gram-negative bacilli, usually affiliated with the pseudomonads or enterics (above). The symptoms of bacterial disease in plants are described by a number of terms such as spots, blights, soft rots, wilts, and galls. *Pseudomonas* species, as well as species of *Burkholderia* and *Ralstonia* (ex-*Pseudomonas*) are among the most important bacteria that are pathogens of plants. They cause economically significant crop disease and crop loss world-wide.

Bacterial spots of various sizes on stems, leaves, flowers and fruits are usually caused by *Pseudomonas* or *Xanthomonas* species. Bacteria may cause spots by producing toxins that kill cells at the site of infection. Blights are caused by rapidly developing necrosis (dead, discoloured areas) on stems, leaves and flowers. Fire blight in apples and pears, caused by *Erwinia amylovora*, can kill young trees within a single season. Bacterial soft rots occur most commonly in fleshy vegetables such as potatoes or onions or fleshy fruits such as tomatoes and eggplants. The most destructive soft rots are caused by *Erwinia* species that attack fruits and vegetables at the post-harvest stage.

Bacterial vascular wilts mainly affect herbaceous plants. The bacteria invade the vessels of the xylem, where they multiply, interfering with the movement of water and inorganic nutrients and resulting in the wilting and the death of the plants. The bacteria commonly degrade portions of the vessel walls and can even cause the vessels to rupture. Once the walls have ruptured, the bacteria then spread to the adjacent parenchyma tissues, where they continue to multiply. In some bacterial wilts, the bacteria ooze to the surface of the stems or leaves through cracks formed over cavities filled with cellular debris, gums, and bacteria.

More commonly, however, the bacteria do not reach the surface of the plant until the plant has been killed by the disease. Wilts of alfalfa and bean plants are cause by species of *Clavibacter;* bacterial wilt of cucurbits, such as squashes and watermelons, are cause by *Erwinia tracheiphila*; the black rot of crucifers such as cabbage is caused by *Xanthomonas campestris*. The most economically-important wilt of plants is caused by *Pseudomonas solanacearum* which affects 44 genera of plants, including such major crops as bananas, peanuts, tomatoes, potatoes, eggplants and tobacco. This disease occurs worldwide in tropical, subtropical, and warm temperate areas.

Mycoplasmas (discussed above) have been identified in more than 200 plant species and associated with more than 50 plant diseases, many with symptoms of yellowing. Among these plant-pathogens are the spiroplasmas (genus *Spiroplasma*), which are pleomorphic, ovoid or spiral-shaped cells which are motile by means of a rotary or screw-like motion. Intracellular fibrils are thought to be responsible for their movement.

The organisms have been isolated from the fluids of vascular plants and from the gut of insects that feed on these fluids. Some have been cultured on artificial media, including *Spiroplasma citri*, which is isolated from the leaves of citrus plants, where it causes citrus stubborn disease, and from corn plants suffering from corn stunt disease.

A number of other mycoplasma-like organisms (sometimes called MLOs) have been detected in diseased plants by electron microscopy, which has been taken as evidence that these organisms may be more involved in plant disease than previously realised.

The causative agent of a common plant disease, termed crown gall, is *Agrobacterium tumefaciens*. The disease is characterised by large galls or swellings that form on the plant at the site of infection, usually near the soil line. Crown gall is a problem in nurseries, affecting ornamental plants and fruit stock, and it may be a serious disease in grapes. Because of their rôle in the genetic engineering of plants, the molecular biology of these bacteria is intensively studied.

PROCARYOTIC CELLS IN BACTERIA

Most procaryotic cells are very small compared to eucaryotic cells. A typical bacterial cell is about 1 micrometer in diameter or width, while most eucaryotic cells are from 10 to 100 micrometers in diameter. Eucaryotic cells have a much greater volume of cytoplasm and a much lower surface: volume ratio than procaryotic cells.

A typical procaryotic cell is about the size of a eucaryotic mitochondrion. Since procaryotes are too small to be seen except with the aid of a microscope, it is usually not appreciated that they are the most abundant form of life on the planet, both in terms of biomass and total numbers of species. For example, in the sea, procaryotes make up 90 percent of the total combined weight of all organisms.

In a single gram of fertile agricultural soil there may be in excess of 10^9 bacterial cells, outnumbering all eucaryotic cells there by 10,000: 1. About 3,000 distinct species of bacteria and archaea are recognised, but this number is probably less than one percent of all the species in nature. These unknown procaryotes, far in excess of undiscovered or unstudied plants, are a tremendous reserve of genetic material and genetic information in nature that awaits exploitation.

Procaryotes are found in all of the habitats where eucaryotes live, but, as well, in many natural environments considered too extreme or inhospitable for eucaryotic cells. Thus, the outer limits of life on Earth (hottest, coldest, driest, etc.) are usually defined by the existence of procaryotes. Where eucaryotes and procaryotes live together, there may be mutualistic associations between the organisms that allow both to survive or flourish. The organelles of eucaryotes (mitochondria and chloroplasts) are thought to be remnants of Bacteria that invaded, or were captured by, primitive eucaryotes in the evolutionary past. Numerous types of eucaryotic cells that exist today are inhabited by endosymbiotic procaryotes. From a metabolic standpoint, the procaryotes are extraordinarily diverse, and they exhibit several types of metabolism that are rarely or never seen in eucaryotes. For example, the biological processes of nitrogen fixation (conversion of atmospheric nitrogen gas to ammonia) and methanogenesis (production of methane) are metabolically-unique to procaryotes and have an enormous impact on the nitrogen and carbon cycles in nature. Unique mechanisms for energy production and photosynthesis are also seen among the Archaea and Bacteria. The lives of plants and animals are dependent upon the activities of bacterial cells. Bacteria and archaea enter into various types of symbiotic relationships with plants and animals that usually benefit both organisms, although a few bacteria are agents of disease.

The metabolic activities of procaryotes in soil habitats have an enormous impact on soil fertility that can affect agricultural practices and crop yields. In the global environment, procaryotes are absolutely essential to drive the cycles of elements that make up living systems, *i.e.*, the carbon, oxygen, nitrogen and sulphur cycles. The origins of the plant cell chloroplast and plant-type (oxygenic) photosynthesis are found in procaryotes. Most of the earth's atmospheric oxygen may have been produced by free-living bacterial cells. The bacteria fix nitrogen and a substantial amount of CO_2, as well. Bacteria or bacterial products (including their genes) can be used to increase crop yield or plant resistance to disease, or to cure or prevent plant disease. Bacterial products include antibiotics to fight infectious disease, as well as components for vaccines used to prevent infectious disease. Because of their simplicity and our relative understanding of their biological processes, the bacteria provide convenient laboratory models for study of the molecular biology, genetics, and physiology of all types of cells, including plant and animal cells.

STRUCTURE AND FUNCTION OF PROCARYOTIC CELLS

Procaryotic cells have three architectural regions: appendages (proteins attached to the cell surface) in the form of flagella and pili; a cell envelope consisting of a capsule, cell wall and plasma membrane; and a cytoplasmic region that contains the cell genome (DNA) and ribosomes and various sorts of inclusions.

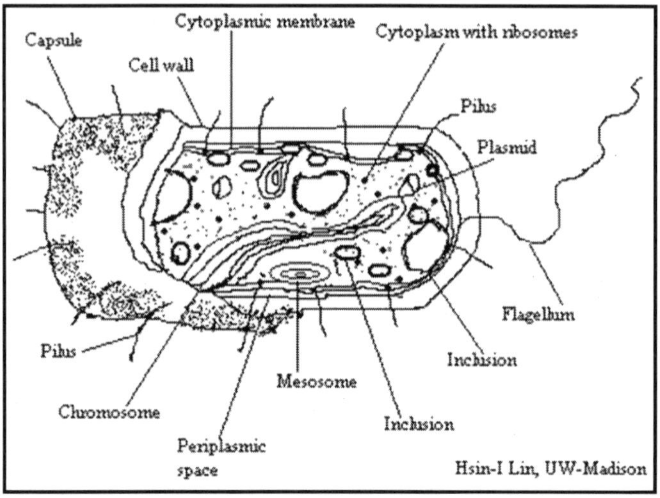

Fig. Schematic Drawing of a Typical Bacterium.

SURFACE STRUCTURES-APPENDAGES

Flagella are filamentous protein structures attached to the cell surface that provide swimming movement for most motile procaryotic cells. The flagellar filament is rotated by a motor apparatus in the plasma membrane allowing the cell to swim in fluid environments. Bacterial flagella are powered by proton motive force (chemiosmotic potential) established on the bacterial membrane, rather than ATP hydrolysis which powers eucaryotic flagella and cilia. Procaryotes are known to exhibit a variety of types of tactic behaviour, *i.e.*, the ability to move (swim) in response to environmental stimuli. For example, during chemotaxis a bacterium can sense the quality and quantity of certain chemicals in their environment and swim towards them (if they are useful nutrients) or away from them (if they are harmful substances).

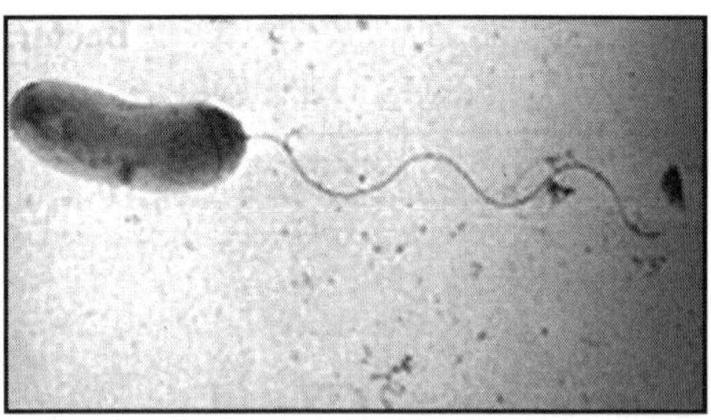

Fig. *Vibrio Cholerae* has a Single Polar Flagellum for Swimming Movement. Electron Micrograph of *Vibrio Cholerae* by Leodotia Pope.

Fimbriae and Pili are interchangeable terms used to designate short, hair-like structures on the surfaces of procaryotic cells. Fimbriae are shorter and stiffer than flagella, and slightly smaller in diameter. Like flagella, they are composed of protein. A specialised type of pilus, the F or sex pilus, in some way stabilises the transfer of DNA between mating bacteria, but the function of the smaller, more numerous common pili is quite different. Common pili (often called fimbriae) are usually involved in adherence (attachment) of procaryotes to surfaces in nature. In medical situations, they are major determinants of bacterial virulence because they allow pathogens to attach to (colonise) tissues and to resist attack by phagocytic white blood cells.

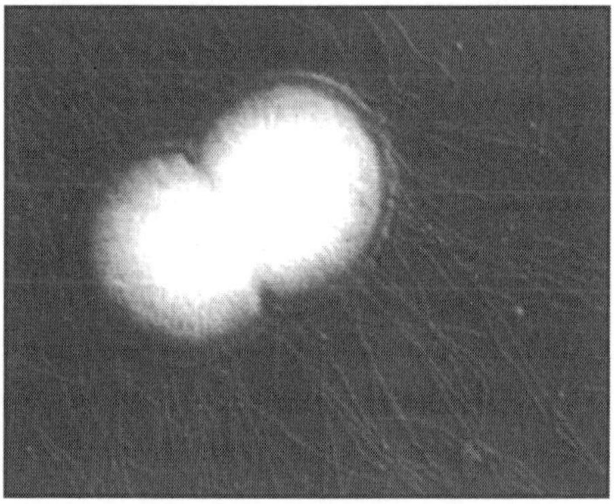

Fig. Fimbriae of *Neisseria Gonorrhoeae* Allow the Bacterium to Adhere to Tissues. Electron Micrograph by David M. Phillips, Visuals Unlimited, with Permission.

RIGID CELL WALL

Most procaryotes have a rigid cell wall. The cell wall is an essential structure that protects the delicate cell protoplast from osmotic lysis. The cell wall of Bacteria consists of a polymer of disaccharides cross-linked by short chains of amino acids (peptides). This molecule is a type of peptidoglycan, which is called murein. In the Gram-positive bacteria (those that retain the purple crystal violet dye when subjected to the Gram-staining procedure) the cell wall is a thick layer of murein.

In the Gram-negative bacteria (cells which do not retain the crystal violet dye) the cell wall is relatively thin and is composed of a thin layer of murein surrounded by a membranous structure called the outer membrane. Murein is a substance unique in nature to bacterial cell walls. Also, the outer membrane of Gram-negative bacteria invariably contains a unique component, lipopolysaccharide (LPS or endotoxin), which is toxic to animals. The cell walls of Archaea may be composed of protein, polysaccharides, or peptidgolycan-like molecules, but never do they contain murein. This feature distinguishes the Bacteria from the Archaea. Although procaryotes lack any intracellular organelles for respiration or photosynthesis, many species possess the physiologic ability to conduct these processes, usually as a function of their plasma membrane. For example, the electron transport system that couples aerobic respiration and ATP synthesis is found in the plasma membrane. The photosynthetic chromophores that harvest light energy for conversion into chemical energy are located in the membrane. Hence, the plasma membrane is the site of oxidative phosphorylation and photophosphorylation in procaryotes, analogous to the functions of mitochondria and chloroplasts in eucaryotic cells. The procaryotic plasma membrane is also a permeability barrier, and it contains a variety of different transport systems that selectively mediate the passage of substances into and out of the cell.

The membranes of Bacteria are structurally similar to the cell membranes of eucaryotes, except that bacterial membranes consist of saturated or monounsaturated fatty acids (rarely polyunsaturated fatty acids) and do not normally contain sterols. The membranes of Archaea form phospholipid bilayers functionally equivalent to bacterial membranes, but archaeal lipids are saturated, branched, repeating isoprenoid subunits that attach to glycerol via an ether linkage, as opposed to the ester linkage found in glycerides of eucaryotic and bacterial membrane lipids. The structure of archaeal membranes is thought to be an adaptation to their existence in extreme environments.

Most bacteria contain some sort of a polysaccharide layer outside of the cell wall or outer membrane. In a general sense, this layer is called a capsule or glycocalyx, bacteria and archaea may also have an additional proteinaceous coat called an S-layer. Capsules, slime layers, glycocalyx and s-layer are known to mediate attachment of bacterial cells to particular surfaces. Capsules also protect bacteria from engulfment by predatory protozoa or white blood cells (phagocytes) and from attack by antimicrobial agents of plant or animal origin. Capsules in certain soil bacteria protect them from perennial effects of drying or desiccation.

VARIOUS SURFACE COMPONENTS OF A PROCARYOTIC CELL

All of the various surface components of a procaryotic cell are important in its ecology since they mediate the contact of the cell with its environment. The only "sense" that a procaryote has results from its immediate contact with its environment. It must use its surface components to assess the environment and respond in a way that supports its own existence and survival in that environment.

The surface properties of a procaryote are determined by the exact molecular composition of its plasma membrane cell wall, including LPS, and the function of surface structures such as flagella, fimbriae and capsules. Some important ways that procaryotes use their surface components are (1) as permeability barriers that allow selective passage of nutrients and exclusion of harmful substances; (2) as "adhesins" used to attach or adhere to specific surfaces or tissues; (3) for protection against engulfment by phagocytic white blood cells or predatory protozoa: (4) as enzymes to mediate specific reactions on the cell surface important in the survival of the procaryote; (5) as "sensing proteins" that can respond to temperature, osmolarity, salinity, light, oxygen, nutrients, etc., resulting in a signal to the genome of the cell that will cause a biological response to a changing environment.

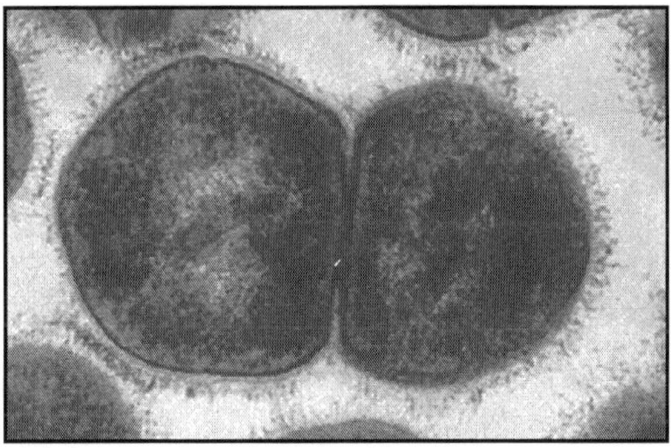

Fig. The Complex Surface of *Streptococcus Pyogenes.* Electron Micrograph of *Streptococcus Pyogenes* by Maria Fazio and Vincent A. Fischetti, Ph.D. with Permission. The Laboratory of Bacterial Pathogenesis and Immunology, Rockefeller University.

CYTOPLASMIC CONSTITUENTS

The cytoplasmic constituents of bacteria invariably include the procaryotic chromosome and ribosomes. The chromosome is typically one large circular molecule of DNA, more or less free in the cytoplasm, although intermittently associated with membranes.

Procaryotes sometimes possess smaller extrachromosomal pieces of DNA called plasmids. The total DNA content of a cell is referred to as the cell genome. During cell growth and division, the procaryotic chromosome is replicated in the usual semi-conservative fashion before for distribution to progeny cells. However, the eucaryotic processes of meiosis and mitosis are absent in procaryotes. Replication and segregation of procaryotic DNA is coordinated by the plasma membrane.

The distinct granular appearance of procaryotic cytoplasm is due to the presence and distribution of ribosomes The ribosomes of procaryotes are smaller than cytoplasmic ribosomes of eucaryotes. Procaryotic ribosomes are 70S in size, being composed of 30S and 50S subunits.

The 80S ribosomes of eucaryotes are made up of 40S and 60S subunits. Ribosomes are involved in the process of translation (protein synthesis), but some details of their activities differ in eucaryotes, Bacteria and Archaea. Protein synthesis using bacterial 70S ribosomes occurs in eucaryotic mitochondria and chloroplasts, and this is taken as a major line of evidence that these organelles are descended from bacteria.

Often contained in the cytoplasm of procaryotic cells is one or another of some type of inclusion granule. Inclusions are distinct granules that may occupy a substantial part of the cytoplasm. Inclusion granules are usually reserve materials of some sort. For example, carbon and energy reserves may be stored as glycogen (a polymer of glucose) or as polybetahydroxybutyric acid (a type of fat) granules. Polyphosphate inclusions are reserves of PO_4 and possibly energy; elemental sulphur (sulphur globules) are stored by some phototrophic and some lithotrophic procaryotes as reserves of energy or electrons. Some inclusion bodies are actually membranous vesicles or intrusions into the cytoplasm which contain photosynthetic pigments or specialised enzyme complexes.

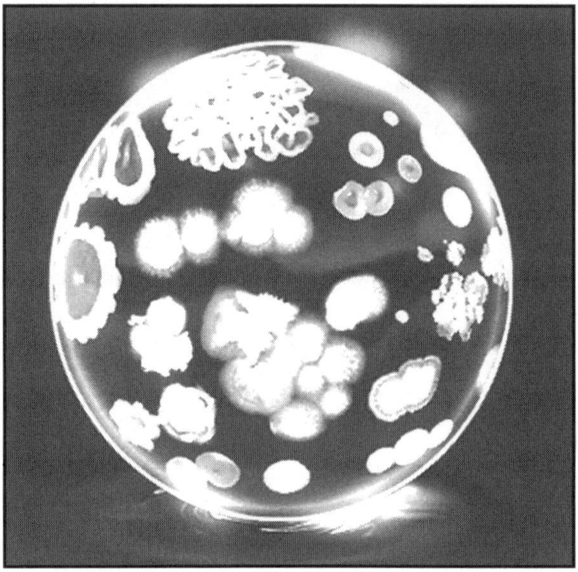

Fig. Bacterial Colonies Growing in a Petri Dish Containing Nutrients. Hans Knoll Institute, Jena, Germany.

TAXONOMY AND CLASSIFICATION OF PROCARYOTES

Haeckel (1866) was the first to create a natural Kingdom for the microorganisms, which had been discovered nearly two centuries before by van Leeuwenhoek. He placed all unicellular (microscopic) organisms in a new kingdom, "Protista", separated from plants (Plantae) and animals (Animalia), which were multicellular (macroscopic) organisms. The development of the electron microscope in the 1950's revealed a fundamental dichotomy among Haeckel's "Protista": some cells contained a membrane-enclosed nucleus, and some cells lacked this intracellular structure.

The latter were temporarily shifted to a fourth kingdom, Monera (or Moneres), the procaryotes (also called Procaryotae). Protista remained as a kingdom of unicellular eucaryotic microorganisms. Whittaker refined the system into five kingdoms in 1967, by identifying the Fungi as a separate multicellular eucaryotic kingdom of organisms, distinguished by their absorptive mode of heterotrophic nutrition.

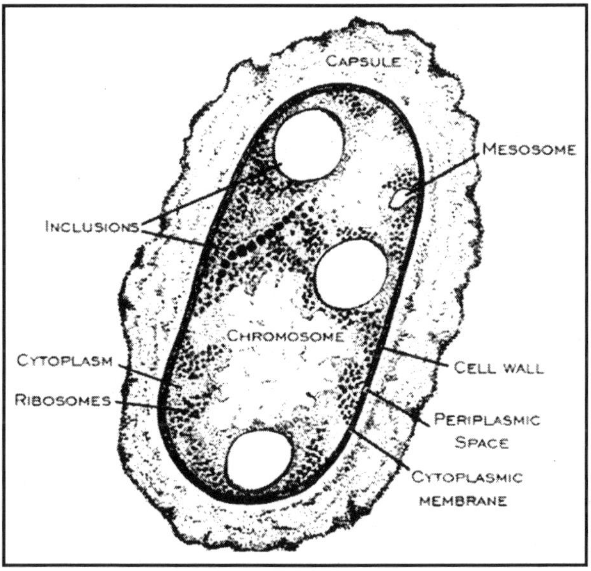

Fig. (above) The Structure of a Typical Procaryotic Cell, in this Case, a Gram-negative Bacterium, Compared with (Below) a Typical Eucaryotic Cell (Plant Cell). The Procaryote is about 1 Micrometer in Diameter and about the size of the Eucaryotic Chloroplast or Mitochondrion. Drawings by Vaike Haas, University of Wisconsin-Madison.

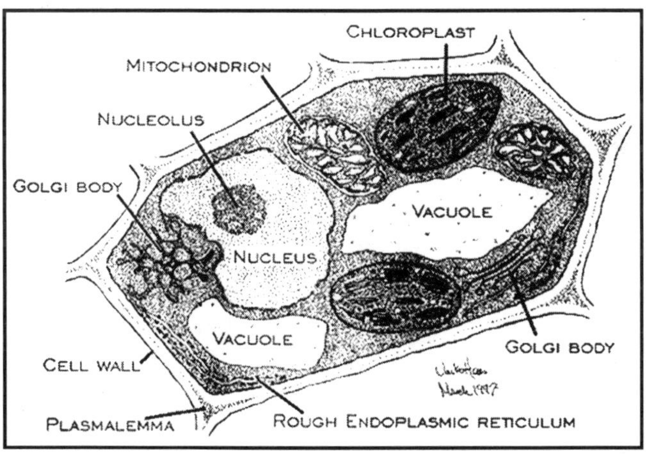

In the 1980's, Woese began phylogenetic analysis of all forms of cellular life based on comparative sequencing of the small subunit ribosomal RNA (ssrRNA) that is contained in all organisms. A new dichotomy was revealed, this time among the procaryotes: there existed two types of procaryotes, as fundamentally unrelated to one another as they are to eucaryotes. Thus, Woese defined the three cellular Domains of life: Eucarya, Bacteria and Archaea. Whittaker's Plant, Animal and Fungal Kingdoms (all of the multicellular eucaryotes) are at the end of a very small branch of the tree of life, and all other branches lead to microorganisms, either procaryotes (Bacteria and

Archaea), or protists (unicellular algae and protozoa), thus establishing clearly that microbial life is the predominant form of life on the planet. Sequence analysis of macromolecules such as the small subunit ribosomal RNA found in all cells has allowed bacteriologists to classify bacteria into a typical hierarchal scheme based on genetic relatedness. The current edition (2001) of Bergey's Manual of Systematic Bacteriology has established 24 phyla of bacteria, systematically ordered into class, order, family, genus and species.

For example, *E. coli* is in the Domain Bacteria, Phylum *Proteobacteria*, Class *Gamma Proteobacteria*, Order *Enterobacteriales*, Family *Enterobacteriaceae*, Genus *Escherichia*, Species *E. coli*. Although the definitive difference between Woese's Archaea and Bacteria is based on fundamental differences in the nucleotide base sequence in the 16S ribosomal RNA, there are many biochemical and phenotypic differences between the two groups of procaryotes. The phylogenetic tree indicates that Archaea are more closely related to Eucarya than are Bacteria. This relatedness seems most evident in the similarities between transcription and translation in the Archaea and the Eucarya. However, it is also evident that the Bacteria have evolved into chloroplasts and mitochondria, so that these eucaryotic organelles derive their lineage from this group of procaryotes. Perhaps the biological success of eucaryotic cells springs from the evolutionary merger of the two procaryotic life forms.

Table. Phenotypic Properties of Bacteria and Archaea Compared with Eucarya.

Property	Biological Domain		
	Eucarya	**Bacteria**	**Archaea**
Cell configuration	eucaryotic	procaryotic	procaryotic
Nuclear membrane	present	absent	absent
Number of chromosomes	>1	1	1
Chromosome topology	linear	circular	circular
Murein in cell wall	-	+	-
Cell membrane lipids	ester-linked glycerides; unbranched; polyunsaturated	ester-linked glycerides; unbranched; saturated or monounsaturated	ether-linked branched; saturated
Cell membrane sterols	present	absent	absent
Organelles (mitochondria and chloroplasts)	present	absent	absent
Ribosome size	80S (cytoplasmic)	70S	70S
Cytoplasmic streaming	+	-	-
Meiosis and mitosis	present	absent	absent
Transcription and translation coupled	-	+	+
Amino acid initiating protein synthesis	methionine	N-formyl methionine	methionine
Protein synthesis inhibited by streptomycin and chloramphenicol	-	+	-
Protein synthesis inhibited by diphtheria toxin	+	-	+

IDENTIFICATION OF BACTERIA CLASSIC METHODS

The criteria used for microscopic identification of procaryotes include cell shape and grouping, Gram-stain reaction, and motility. Bacterial cells almost invariably take one of three forms: rod (bacillus), sphere (coccus), or spiral (spirilla and spirochetes). Rods that are curved are called vibrios. Fixed bacterial cells stain either Gram-positive (purple) or Gram-negative (pink); motility is easily determined by observing living specimens. Bacilli may occur singly or form chains of cells; cocci may form chains (streptococci) or grape-like clusters (staphylococci); spiral shape cells are almost always motile; cocci are almost never motile. This nomenclature ignores the actinomycetes, a prominent group of branched bacteria which occur in the soil. But they are easily recognised by their colonies and their microscopic appearance.

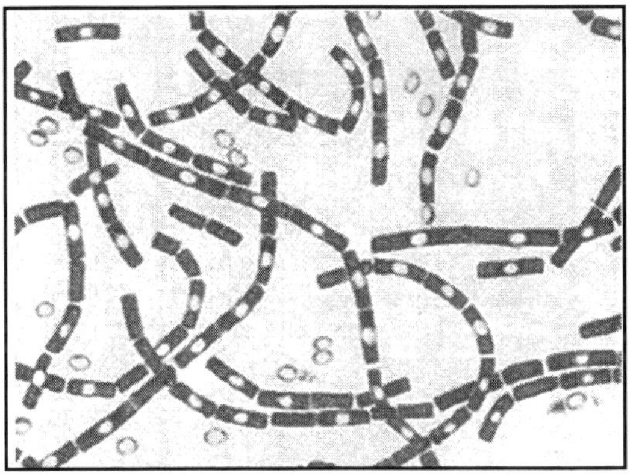

Fig. Gram Stain of *Bacillus Anthracis, the* Cause of Anthrax. K. Todar.

Such easily-made microscopic observations, combined with knowing the natural environment of the organism, are important aids to identify the group, if not the exact genus, of a bacterium - providing, of course, that one has an effective key. Bergey's Manual describes affiliated groups of Bacteria and Archaea based on a few easily observed microscopic and physiologic characteristics. Further identification requires biochemical tests which will distinguish genera among families and species among genera. Strains within a single species are usually distinguished by genetic or immunological criteria.

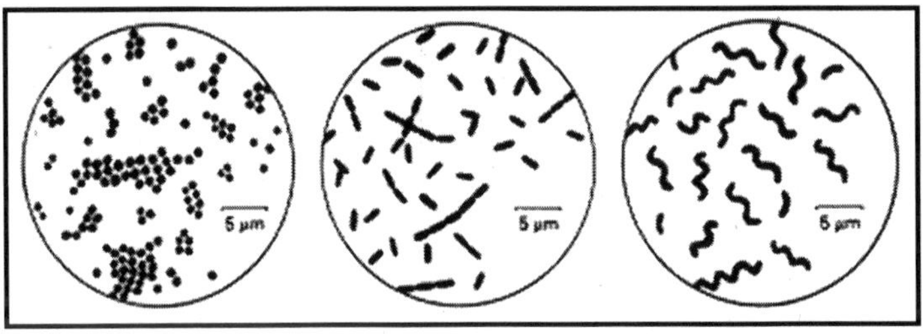

Fig. Size and Fundamental Shapes of Procaryotes Revealed by three Genera of Bacteria (l to r): *Staphylococcus* (Spheres), *Lactobacillus* (Rods), and *Aquaspirillum* (spirals).

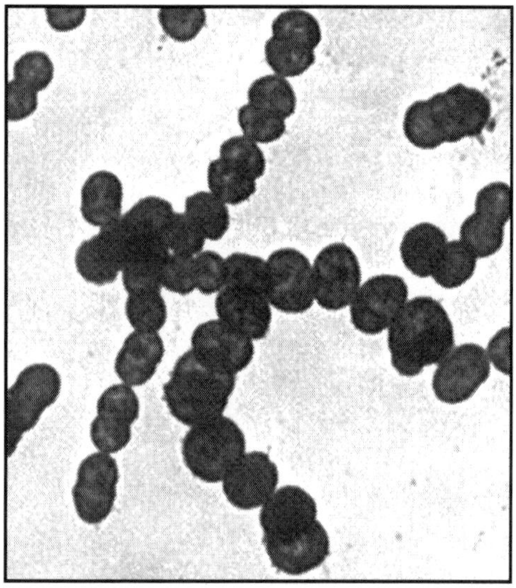

Fig. Streptococcus Pyogenes by Maria Fazio and Vincent A. Fischetti, Ph.D. with Permission. The Laboratory of Bacterial Pathogenesis and Immunology, Rockefeller University.

MOLECULAR TECHNIQUES

The sciences of genomics and bioinformatics have led to a radical reclassification of procaryotes based on comparative analysis of organismal DNA. Genomics involves the study of all of the nucleotide sequences, including structural genes, regulatory sequences, and non-coding DNA segments, in the chromosomes of an organism. To date over 200 bacterial genomes have been sequenced.

We have seen how highly conserved genetic sequences, such as those that encode for the small subunit ribosomal RNAs (16S rRNA) of an organism, can be analysed to specifically relate two organisms. So can the identification of certain genes provide information about specific properties of an organism, and analysis of specific nucleotide sequences may be used to indicate identity and degrees of genetic relatedness among organisms.

The newest editions of Bergey's Manual are adapted to the new phylogenetic classification. This has resulted in the formation of several new taxa of bacteria and archaea at every hierarchical level. Occasionally, organisms thought to be more or less distantly related become unified; but more likely, organisms thought to be closely-related due to similar phenotypic properties are found to be genetically distinct and warrant separation into a new taxa.

Metagenomics. Sequencing of 16S rRNA genes obtained from environmental samples produces a broad profile of microbial diversity and reveals that the vast majority of microbes present have been missed by reliance on cultivation-based methods. This observation has given rise to the field of metagenomics. Metagenomics (also called environmental genomics) is the application of modern genomics techniques to the study of communities of microorganisms directly in their natural environments, bypassing the need for isolation and lab cultivation of individual species.

Metagenomics provides a means to identify and quantify microbes from environmental samples based on the presence of distinctive genes. This enables studies of organisms that are not easily cultured in the laboratory, as well as studies of organisms in their natural environment.

"Shotgun" metagenomics is capable of sequencing nearly complete microbial genomes directly from the environment. Because the collection of DNA from an environment is largely uncontrolled, the most abundant organisms in a sample are most highly represented in the resulting sequence data.

To achieve the high coverage needed to fully resolve the genomes of underrepresented community members, large samples, often prohibitively so, may be needed. However, the random nature of shotgun sequencing ensures that many of these organisms will be represented by at least some small sequence segments. Due to the limitations of microbial isolation methods, the majority of these organisms would go unnoticed using traditional culturing techniques. Shotgun sequencing and screens of clone libraries reveal genes present in environmental samples. This provides information both on which organisms are present and what metabolic processes are possible in the community. This can be helpful in understanding the ecology of a community, particularly if multiple samples are compared to one other.

BACTERIAL REPRODUCTION AND GENETICS

Most bacteria reproduce by a relatively simple asexual process called binary fission: each cell increases in size and divides into two cells. During this process there is an orderly increase in cellular structures and components, replication and segregation of the bacterial DNA, followed by formation of a septum or cross wall which divides the cell into two.

The process is evidently coordinated by activities associated with the cell membrane. The DNA molecule is believed to be attached to a point on the membrane where it is replicated. The two DNA molecules remain attached at points side-by-side on the membrane while new membrane material is synthesized between the two points.

This draws the DNA molecules in opposite directions while new cell wall and membrane are laid down as a septum between the two chromosomal compartments. When septum formation is complete the cell splits into two progeny cells. The time interval required for a bacterial cell to divide or for a population of cells to double is called the generation time. Generation times for bacterial species growing in nature may be as short as 15 minutes or as long as several days.

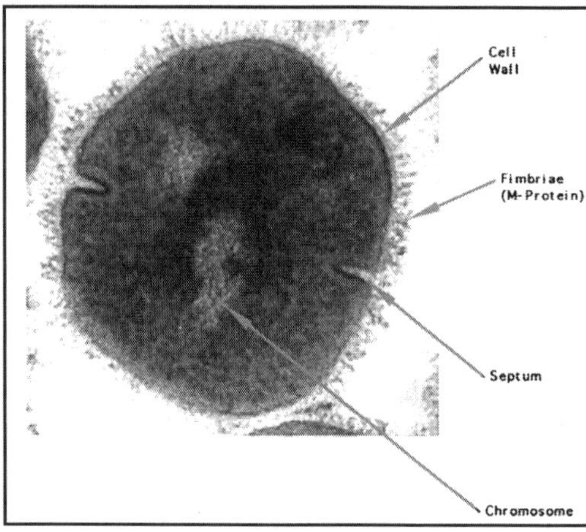

Fig. A Pair of Dividing Streptococci. The Chromosome has been Replicated and is Partially Segregated as Septum Formation is Beginning. Electron Micrograph of *Streptococcus Pyogenes* by Maria Fazio and Vincent A. Fischetti, Ph.D. with permission.

GENETIC EXCHANGE IN BACTERIA

Although procaryotes do not undergo sexual reproduction, they are not without the ability to exchange genes and undergo genetic recombination. Bacteria are known to exchange genes in nature by three fundamental processes: conjugation, transduction and transformation. Conjugation requires cell-to-cell contact for DNA to be transferred from a donor to a recipient. During transduction, a virus transfers the genes between mating bacteria. In transformation, DNA is acquired directly from the environment, having been released from another cell. Genetic recombination can follow the transfer of DNA from one cell to another leading to the emergence of a new genotype (recombinant). It is common for DNA to be transferred as plasmids between mating bacteria. Since bacteria usually develop their genes for drug resistance on plasmids (called resistance transfer factors, or RTFs), they are able to spread drug resistance to other strains and species during genetic exchange processes. The genetic engineering of bacterial cells in the research or biotechnology laboratory is often based on the use of plasmids as vectors. The genetic systems of the Archaea are poorly characterised at this point, although the entire genome of *Methanosarcina* has been sequenced which opens up the possibilities for genetic analysis of the group.

EVOLUTION OF BACTERIA AND ARCHAEA

For most procaryotes, mutation is a major source of variability that allows the species to adapt to new conditions. The mutation rate for most procaryotic genes is in the neighbourhood of 10^{-8}. This means that if a bacterial population doubles from 10^8 cells to 2×10^8 cells, there is likely to be a mutant present for any given gene. Since procaryotes grow to reach population densities far in excess of 10^9 cells, such a mutant could develop from a single generation during

15 minutes of growth. The evolution of procaryotes, driven by such Darwinian principles of evolution (mutation and selection) is called vertical evolution.

However, as a result of the processes of genetic exchange described above, the bacteria and archaea can also undergo a process of horizontal evolution, also called horizontal gene transfer (HGT). In this case, genes are transferred laterally from one organism to another, including between members of different Kingdoms, which allows the recipient to experiment with a new genetic trait. Horizontal gene transfer is becoming realised to be a significant force in driving cellular evolution.

The combined effects of fast growth rates, high concentrations of cells, genetic processes of mutation and selection, and the ability to exchange genes, account for the extraordinary rates of adaptation and evolution that can be observed in the procaryotes.

ECOLOGY OF BACTERIA AND ARCHAEA

Bacteria and Archaea are present in all environments that support life. They may be free-living, or living in associations with "higher forms" of life (plants and animals), and they are found in environments that support no other form of life. Procaryotes have the usual nutritional requirements for growth of cells, but many of the ways that they utilise and transform their nutrients are unique. This bears directly on their habitat and their ecology.

NUTRITIONAL TYPES OF ORGANISMS

In terms of carbon utilisation a cell may be heterotrophic or autotrophic. Heterotrophs obtain their carbon and energy for growth from organic compounds in nature. Autotrophs use CO_2 as a sole source of carbon for growth and obtain their energy from light (*e.g.*, photoautotrophs) or from the oxidation of inorganic compounds (*e.g.*, lithoautotrophs).

Most heterotrophic bacteria are saprophytes, meaning that they obtain their nourishment from dead organic matter. In the soil, saprophytic bacteria and fungi are responsible for biodegradation of organic material. Ultimately, organic molecules, no matter how complex, can be degraded to CO_2 (plus H_2 and H_2O). Probably no naturally-occurring organic substance cannot be degraded by the combined activities of the bacteria and fungi. Hence, most organic matter in nature is converted by heterotrophs to CO_2, only to be converted back into organic material by autotrophs that die and nourish heterotrophs to complete the carbon cycle. Lithotrophic procaryotes have a type of energy-producing metabolism which is unique. Lithotrophs (also called lithoautotrophs or chemoautotrophs) use inorganic compounds as sources of energy, *i.e.*, they oxidise compounds such as H_2 or H_2S or NH_3 to obtain electrons to feed in to an electron transport system and to produce ATP. Lithotrophs are found in soil and aquatic environments wherever their energy source is present. Most lithotrophs are autotrophs so they can grow in the absence of any organic material. Lithotrophic species are found among the Bacteria and the Archaea. Sulphur-oxidising lithotrophs convert H_2S to S^o and S^o to SO_4. Nitrifying bacteria convert NH_3 to NO_2 and NO_2 to NO_3; methanogenic archaea strip electrons off of H_2 as a source of energy and add them to CO_2 to form CH_4 (methane). Lithotrophs have an obvious impact on the sulphur, nitrogen and carbon cycles in the biosphere.

Photosynthetic bacteria convert light energy into chemical energy for growth. Most phototrophic bacteria are autotrophs so their role in the carbon cycle is analogous to that of plants. The planktonic cyanobacteria are the "grass of the sea" and their form of oxygenic photosynthesis generates a substantial amount of O_2 in the biosphere. However, among the photosynthetic bacteria are types of photosynthetic metabolism not seen in eucaryotes, including photoheterotrophy (using light as an energy source while assimilating organic compounds as a source of carbon), anoxygenic photosynthesis, and unique mechanisms of CO_2 fixation (autotrophy).

Photosynthesis has not been found to occur among the Archaea, but one archaeal species employs a light-driven non- photosynthetic means of energy generation based on the use of a chromophore called bacteriorhodopsin.

ADAPTATIONS TO ENVIRONMENTAL CONDITIONS

Most procaryotes, whether they have been cultured and studied in the laboratory, or observed growing in their natural habitats, seem to be highly adapted to their specific environment by means of their macromolecular structure and/or their physiologic (metabolic) capabilities. The nutritional quality of the environment determines whether a particular organism will be present, but so do various physical parameters such as the availability of light and O_2, as well as the pH, temperature and salinity of the environment. As examples, the range of procaryotic responses to oxygen and temperature are discussed below.

Procaryotes vary widely in their response to O_2 (molecular oxygen). Organisms that require O_2 for growth are called obligate aerobes; those which are inhibited or killed by O_2, and which grow only in its absence, are called obligate anaerobes; organisms which grow either in the presence or absence of O_2 are called facultative anaerobes. Whether or not a particular organism can exist in the presence of O_2 depends upon the distribution of certain enzymes such as superoxide dismutase and catalase that are required to detoxify lethal oxygen radicals that are always generated by living systems in the presence of O_2 Procaryotes also vary widely in their response to temperature. Those that live at very cold temperatures (0 degrees or lower) are called psychrophiles; those which flourish at room temperature (25 degrees) or at the temperature of warm-blooded animals (37 degrees) are called mesophiles; those that live at high temperatures (greater than 45 degrees) are thermophiles. The only limit that seems to be placed on growth of certain procaryotes in nature relative to temperature is whether liquid water exists. Hence, growing procaryotic cells can be found in supercooled environments (ice does not form) as low as -20 degrees and superheated environments (steam does not form) as high as 120 degrees. Archaea have been detected around thermal vents on the ocean floor where the temperature is as high as 320 degrees!

SYMBIOSIS

The biomass of procaryotic cells in the biosphere, their metabolic diversity, and their persistence in all habitats that support life, ensures that these microbe will play a crucial role in the cycles of elements and the functioning of the world ecosystem. However, the procaryotes affect the world ecology in another significant way through their inevitable interactions with insects, plants and animals.

Some bacteria are required to associate with insects, animals or plants for the latter to survive. For example, the sex of offspring of certain insects is determined by endosymbiotic bacteria. Ruminant animals (cows, sheep, etc.), whose diet is mainly cellulose (plant material), must have cellulose-digesting bacteria in their intestine to convert the cellulose to a form of carbon that the animal can assimilate. Leguminous plants grow poorly in nitrogen-deprived soils unless they are colonised by nitrogen-fixing bacteria which can supply them with a biologically-useful form of nitrogen.

6

Viral an Infectious Disease

INTRODUCTION

Zoonosis (zoo-e-no-sis) is an infectious disease that may be transmitted from animals (wild and domestic) to humans or from humans to animals. The word zoonosis is derived from the Greek, *zoon* (animal) (pronounced as zoo-on) and *nosos* (disease). Of the 1415 microbial diseases affecting humans, 61 per cent are zoonotic and among emerging infectious diseases, 75 per cent are zoonotic with wildlife being one of the major sources of infection. A new virus has been emerging almost every year since last two decades. Of 534 zoonotic viruses (belonging to 8 families) identified 120 cause human illnesses with or without the involvement of intermediate host/vectors.

In the past 15 years, many zoonotic viral infections are of emerging and re-emerging in nature and haemorrhagic fever causing viruses transmitted by insect vectors (arboviruses *i.e.*, yellow fever virus), rodents *i.e.*, Hanta viruses and also by direct contact *i.e.*, Filoviruses.

Thus, they pose a great challenge to both veterinary and public health professionals. It is essential to investigate the complex interactions between pathogens, host, vectors and environment to curtail these infections. This review focuses on description of the important zoonotic viral infections with especially the recently emerging and reemerging diseases and their causes, transmission, clinical manifestations, distribution and preventive measures, to abreast the knowledge on zoonoses.

Transmission Zoonotic viruses are transmitted to humans either directly or indirectly. Direct transmission involves contact between the infected and susceptible individual (orf), bite (rabies) and handling of the affected animal's tissues or materials (Orf).

Indirect transmission involves transmission through the bite of a hematophagous (blood-sucking) arthropod after replicating in the reservoir animal host (Japanese encephalitis, yellow fever). Most viral zoonoses require blood-sucking arthropods for their transmission to humans. Among them, mosquitoes (Equine encephalitis complex) are the most common followed by ticks (Powassan virus), sand flies (Vesicular stomatitis) and midges (bluetongue). The arthropod vector becomes infected when it feeds the blood of a viraemic animal. In most of the cases, virus replicates in the arthropod tissues and reaches their salivary glands. The arthropod then transmits the virus to a new susceptible host when it injects infective salivary fluid while taking a blood meal. The extrinsic incubation period (time between ingestion and transmission of the virus) is usually 8 to 12 days. This period depends on the virus, the environment and the vector species involved.

Arthropod-borne viruses generally remain undetected until humans encroach on the natural enzootic focus or until the virus escapes the primary cycle via a secondary vector or vertebrate host. Wild birds are important to public health as they carry various zoonotic pathogens and they either act as reservoir hosts or help in disseminating the infected arthropod vectors. In addition, bird migration provides a mechanism for the establishment of new endemic foci of disease at great distances from where an infection was acquired (avian influenza).

There has been a change in the transmission pattern especially in the occurrence and incidence of diseases due to broadening of host range (Monkey pox and Nipah viruses), high mutation rate (avian influenza, FMD) and anthropogenic environmental changes *viz.*, ecological imbalance and change in agricultural practices.

ROLE OF WILDLIFE IN ZOONOSIS

The significance of wild life as animal reservoir for zoonotic viruses has been traced long back with two important ancient diseases such as rabies and West Nile virus and represent as large spectrum of transmission mode. Of the total emerging diseases, 75 per cent are considered zoonotic with wild life as a major source of reservoir.

Recent emerging viral diseases which moved into new species such as AIDS, SARS and avian influenza have a strong evidence of wild life origin due to human encroachment and changed international trade and travel patterns. Commonly the pattern of moving of viral agents from wild animal species to human occurs either as actual transmission being rare (HIV, Influenza A, Ebola and SARS) but will be maintained and has potential of man to man transmission or direct/indirect manner through animal bite and arthropod vectors (rabies, Nipah, West Nile virus and hantavirus). Many zoonoses with a wildlife origin are spread through insect vectors (Rift Valley fever, equine encephalitis and Japanese encephalitis), whereas, rabies by animal bite and hantaviruses by contact with rodent excreta is common.

The outcome in the form of clinical manifestation in humans depends on the transmission pattern of the agent causing the disease. Direct contact and vector bite lead to the formation of rashes and ulcers, whereas, intake of contaminated meat/water lead to digestive tract problems and diseases transmitted by inhalation of infected foci of dust cause pneumonia like illness. Wild life are basically involved in epidemiology of the disease which is influenced by other factors such as change in agro-climatic conditions, host abundance, movement of pathogens/vector/animal host including migratory birds and anthropogenic factors.

For example, increase in transmission and subsequent spread of Sin Nombre Hantavirus causing Hantavirus Pulmonary Syndrome (HPS) to humans is due to increase in heavy rainfall and host abundance in USA. Increase in the emergence of some wild life diseases result in high potential of emergence of human pathogens as in the case of West Nile virus spread in USA.

A potential threat to human health, animal welfare and species conservation from domesticated and wild life is presented equally by emergence of human and wild life pathogens.

MANIFESTATIONS OF VIRAL ZOONOSES

Zoonotic infections are broadly grouped in to:

- Diseases causing no illness,
- Non-specific viral syndrome and
- Severe illness.

The third category of infections is further classified in to:

- Hemorrhagic fever,
- Encephalitis and/or rash arthralgia,
- Emerging and reemerging and
- Rare zoonotic infections.

Encephalitis

The major viral zoonoses, which are associated with encephalitis. They are arthropod borne and belong mostly to five viral families (Rhabdoviridae, Flaviviridae, Togaviridae, Reoviridae and Bunyaviridae). Most of them are transmitted through mosquito or tick bites, except a few which are transmitted through bite of an infected host (rabies). Mosquitoes and ticks are major vectors for this category of infections.

They cause symptoms like fever, vomition, encephalitis, headache and neurological disorders. Some of these infections are confined to a particular country (Colorado tick fever), while others are distributed worldwide (rabies). Prophylactic/therapeutic measures are available for some of the infections, while for others vector elimination is the only means of control.

Intense research is required towards the development of vaccines including conventional as well as recombinant. Specific diagnosis of this group of infections is done employing serological tests like Hemagglutination-Inhibition (HI), Complement Fixation (CF) and Virus Neutralisation (VN).

Hemorrhagic Fevers

Most of the viral zoonoses causing haemorrhagic fevers are reported to be of emerging and reemerging in nature. There are more than 16 zoonotic infections in this category belong mainly to four viral families (Arenaviridae, Bunyaviridae, Flaviviridae, Filoviridae). These infections are often associated with extensive bleeding in human. Most of them are transmitted upon vector bite. The common vectors are mosquitoes and ticks. Vaccines are not available for majority of the infections and therefore, control relies on supportive treatment. Control of vector is the main means of control. Chemotherapy is available for some of the infections (Crimean-Congo haemorrhagic fever) with a limited success.

Rashes and Arthralgia

A very few viruses are associated with local rashes and arthralgia and almost all belong to *Togaviridae* family. Most of them are transmitted to humans through infected mosquito bites. These vectors are mainly from *Aedes* and *Culex* families. No specific treatment is available and control depends on the elimination of vectors. EU countries appear to be free, while other continents are endemic for these infections.

Emerging and Reemerging Zoonoses

The complex interaction between environment/ecology, social, health care, human demographics and behaviour influences the emergence and re-emergence of zoonotic viral diseases. Periodic discovery of new zoonoses suggest that the known viruses are only a fraction of the total number that exist in nature. The RNA viruses are capable of adapting to changing environmental conditions rapidly and are among the most prominent emerging pathogens. Mutations are more common in RNA viruses (Influenza) than DNA viruses (Pox).

The common mutations are point (insertion/deletion), drift (minor) and shift (major). In addition to these, movement of population, birds, vectors, pathogens and trade contribute to the global spread of emerging infectious diseases (influenza, severe acute respiratory syndrome). Other factors *viz.*, human migration, change in land use pattern, mining (disturbance of ecosystem), coastal land degradation, wetland modification, construction of buildings, habitat fragmentation, deforestation, expansion of agents host range, human intervention in wild life resources like hiking, camping and hunting also influence on acquiring zoonotic infections from wildlife. Cessation of vaccination against smallpox since 1980s, emergence of some genetically related orthopoxviruses has been reported throughout the world *i.e.*, monkey pox, buffalo pox and Bovine Vaccinia (BV) infections. Despite successful eradication of some viral diseases (small pox and almost polio in humans and rinderpest in cattle) due to intensive research and dedicated coordinated efforts, modern medicine has failed to control many infectious diseases resulting from emerging and reemerging viruses. Some infectious agents already known to be pathogenic have gained increasing importance in recent decades due to change in disease patterns. Several previously unknown infectious agents with a high pathogenic potential have also been identified. Several infectious viral agents (DNA and RNA viral families) have been emerged as zoonotic agents. They are associated with flu-like signs (Alkhumra virus infection, influenza A) to respiratory (SARS), pox lesions mostly localised distributed over hairless parts of body namely udder, teats, ears and tail (in buffaloes) and fingers and hands (in humans) due to buffalopox and Orf virus infections in affected goats, hepatitis (hepatitis E virus), haemorrhagic fevers (Ebola, Marburg and hanta virus infections) and encephalitis (Henipa virus complex). Treatment/prophylaxis is not available to many of these infections. But some of antiviral compounds, which are under trial, are found to be effective.

For example Ebola and Marburg viruses are inhibited invitro by Carbocyclic-3-deazaadenosine, a first compound to cure these virus infections.

Rare Viral Zoonoses

Several viral infections cause non-specific febrile illness in humans and occur rarely. Many of them are animal pathogens, but often they produce non-specific febrile illnesses in humans, though, humans are not the primary hosts. However, there is an increasing trend of occurrence of such infections in recent times. Transmission of these infections have been reported upon direct contact of human objects with infected animal (FMD particularly serotypes O followed by C and rarely A, buffalo pox, Orf), handling of such organisms in the laboratory (bluetongue, Newcastle disease), sexual contact (simian immuno deficiency (SID) virus), bite/ scratch (monkey B virus), vectors (semliki forest virus, African horse sickness and louping ill) and food and water (calici viruses such as swine vesicular exanthema, feline calicivirus and rabbit haemorrhagic disease virus causing vomition and diarrohoea. Recently, animal rotaviruses and Eyach virus related to Colorado tick fever virus and Oropouche fever virus, an arbovirus similar to dengue fever in Trinidad are reported to cause mild infections in humans. Treatment is not available for most of the human infections, while some of them can be treated with nucleoside analogues like acyclovir or gancyclovir.

Other Zoonosis

Prion diseases are caused by scrapie associated prion protien (PrP^{sc}), which are proteinacious infectious agents common in animals and humans. Some of the animal prion diseases are scrapie of sheep, Bovine Spongiform Encephalopathy (BSE) and goats and mink spongiform encephalopathy. Human prion diseases are Creutzfeldt-Jakob Disease (CJD), Kuru, Gertsmann Straussler Schienker Syndrome (GSS) and fatal familial insomnia. The human disease variant (vCJD) is believed to be a zoonotic disease caused by BSE agent and recently an emerging disease as well. The route of transmission of vCJD is not yet fully proven but it is generally transmitted through exposure to food contaminated by the bovine BSE agent. Human prion diseases can be classified as sporadic, hereditary or acquired. Acquired form *i.e.*, vCJD is caused by the transmission of infection from human to human or, as a zoonosis, from cattle to human. Transmission of infectious agents between species through xenotransplantation called xenosis is another way of introducing viruses from animal to human (porcine endogenous retroviruses). No specific treatment and vaccine is available. Prevention is by avoiding consumption of BSE contaminated or half cooked meat.

Prevention, Control Measures and Perspectives

Effective prevention and control measures can be achieved through proper diagnostics and prophylactic aids to curtail further spread in most of zoonotic viral diseases. Improved sanitary conditions such as proper treatment and disposal of human waste, higher standards for public water supplies, improved personal hygiene procedures and sanitary food preparation are vital to strengthen the control measures. A clear understanding of epidemiology of the diseases with wild life as reservoir namely the virulence and transmissibility of many diseases (human monkey pox, Tana pox and Yaba pox) could help in understanding the severity and thereby to take appropriate measures in eradication of such dreadful diseases. Research should focus on molecular biology of these viruses so as to develop diagnostics and prophylactics in a modern way to combat these infections in short time.

To safeguard the public health from pathogens of zoonotic infections, application of skills, knowledge and resources of veterinary public health is essential. It is time to combat viral zoonoses with a combined effort of veterinary and public health specialists. A better understanding of avian migration patterns and their infectious diseases would be useful to forecast disease outbreaks due to emerging zoonotic infections like avian influenza. Further, the control measures for emerging and re-emerging viral pathogens are demanding, as there is population explosion.

Novel, highly sensitive and specific techniques comprising genomics and proteomics along with conventional methods would be useful in the identification of emerging and re-emerging viruses, thereby; therapeutic/prophylactic/ preventive measures would be applied on time. The first line of measure to control any disease is the surveillance. Control and prevention strategies should be designed based on transmission pattern and characteristics of virus, involvement of vectors, environment and epidemiology of the disease. The

European Union (EU) has established a net work termed as Med-Vet-Net to develop a network of excellence for the integration of medical, veterinary and food scientists in order to develop food safety measures and to improve research on the prevention and control of zoonoses, including food-borne diseases.

The network will also consider the concerns of consumers and other stakeholders throughout the food chain. Another system the Hazard Analysis and Critical Control Point (HACCP), which is regulated under FDA and it aims at analysing hazards associated with food and identify preventive and control measures to check spread of food-borne diseases including viral pathogens. Similarly, sanitary and phyto-sanitary measures (SPS) measures, which are set out with WTO are to be strictly followed to have safe food in order to conserve the health of animal, human and plants due to zoonotic agents.

INFECTIONS IN ZOONOTIC

Zoonoses are infections or diseases that can be transmitted directly or indirectly between animals and humans, for instance by consuming contaminated foodstuffs or through contact with infected animals. The severity of these diseases in humans varies from mild symptoms to life-threatening conditions. Research indicates that between one third and one half of all human infectious diseases have a zoonotic origin, that is, are transmitted from animals. About 75 per cent of the new diseases that have affected humans over the past 10 years (such as the West Nile Virus) have originated from animals or products of animal origin. Zoonoses are transmissible between animals and humans in a variety of ways and an infection can also often be transmitted through multiple ways:

Food-borne Zoonotic Diseases

- Food-borne zoonotic pathogens are transmitted through consumption of contaminated food or drinking water. Infectious agents in foodstuffs include bacteria such as *Salmonella* and *Campylobacter*, viruses such as norovirus or hepatitis A virus, and parasites such as *Trichinella*.
- The infectious agent which causes Bovine Spongiform Encephalopathy (BSE) in cattle can also be transmitted to humans through consumption of contaminated meat causing variant Creuzfeldt-Jakob disease. Unlike other food-borne diseases which are spread by microorganisms, BSE is caused by a prion, which is an abnormal form of a protein (known as PrP^c).

Non-Food-Borne Zoonotic Diseases

- By vectors, *i.e.* living organisms that transmit infectious agents from an infected animal to a human or another animal. Vectors are frequently arthropods, such as mosquitoes, ticks, flies, fleas and lice and can transmit diseases such as malaria, West-Nile virus and Lyme disease.

- Through direct contact or close proximity with infected animals. Diseases that are mainly transmissible to other animals or humans in this way include:
- Avian influenza, which is a viral disease occurring in poultry and other birds. Pigs can also be carriers of this virus as well as of other influenza viruses. Avian influenza primarily affects birds, but there have been cases of viruses being transmitted to humans and other animals through close contact with infected birds.
- Q fever is a disease, caused by the Coxiela burnetti bacterium, affecting animals and humans. It has been reported to be present in a wide range of species, including cattle, sheep and goats as well as birds and arthropods. Human infection mainly results from the inhalation of dust contaminated with bacteria from the placenta and birth fluids or faeces from infected animals. Other modes of transmission, such as through contaminated water or the feces of infected arthropods are rare.
- A specific strain of the Meticillin-resistant *Staphylococcus aureus* (MRSA) bacterium (CC398) which can be transmitted through contact with live animals.
- *Salmonella* infections can originate from contact with infected reptiles and amphibians such as pet snakes, iguanas and frogs or their environment.
- Verotoxin-producing *Escherichia coli (E. coli)* can be acquired through contact with infected farm animals.
- These diseases can also be transmitted through the environment, *e.g.* Verotoxin-producing *E. coli* in contaminated swimming water.

TRANSMISSION OF VIRAL ZOONOSIS

Viral zoonosis, the transmission of a virus from its primary vertebrate reservoir species to humans, requires ubiquitous cellular proteins known as receptor proteins. Zoonosis can occur not only through direct transmission, but also through intermediate reservoirs or other environmental factors. The zoonotic viruses can be categorised according to genotype; of the various classes of viruses, the RNA viruses exhibit the highest mutation rates. Most zoonotic viruses are RNA viruses that change their envelope proteins to facilitate binding to various receptors of host species. The high mutation rate of envelope proteins hinders the development of accurate vaccines, as does the great ability of the RNA viruses to infect host species in order to exploit host proteins for viral reproduction.

Lacking the ability to self-replicate, viruses must utilise the replication apparatus of their host cells. Viral infection of a cell begins with attachment of the virus to the cell surface. During attachment to the cell membrane, the viral envelope protein (a structural protein) interacts with the host-cell receptor protein(s). In non-envelope viruses, the capsid plays this role. The cell receptors that play a major role in viral attachment are predominantly membrane proteins of the immunoglobin superfamily.

The identification of virus-binding cellular receptors was rapidly accelerated in the late 1980s owing to developments in the use of monoclonal antibodies and molecular cloning techniques. The various receptors that have been found are surface matrix structures containing carbohydrate, lipid, and protein moieties.

In some cases, viral attachment also exploits co-receptors. For example, HIV, which uses the CD4 molecule as its receptor, uses the CXCR4 and CCR5 co-receptors to strengthen the effectiveness of infection. Similarly, hepatitis C virus utilises CD81 as a receptor and LDLR as a co-receptor.

We hypothesised that the major barrier to the transmission of viruses between species is the difference in cellular receptor sequences. In other words, the specific amino acid sequence of the receptor should be the major determinant of the ability of the viral envelope protein to attach to the cell. Ordinary sequence alignment protocol tells us overall sequence similarity which we thought useful but insufficient because most receptors are membrane proteins and membrane proteins consist of distinctive hydrophobic and hydrophilic parts.

Therefore, we analysed host-cell receptor sequences for their hydrophobicity/hydrophilicity characteristics. We then analysed these properties for similarities among receptors of different species to predict the likelihood of transmission across species, including humans. To our best knowledge, this study is the first attempt to predict zoonosis through a simple analysis of receptor sequence similarities and differences. This method may be useful in predicting the zoonotic potential of newly discovered viral strains.

Results and Discussion

The pair-wise receptor sequence similarities (${}^{g}S_{i,1}$, ${}^{g}S_{i,2}$, and ${}^{g}S_{i,3}$) between host-species pairs for each virus family. For logical comparisons, each virus contains at least one infected host. The similarity scores for the infected group (g = 1) were high, ranging from 0.790 to 0.988 for ${}^{1}S_{i,1}$, from 0.841 to 0.996 for ${}^{1}S_{i,2}$, and 0.794 to 0.962 for ${}^{1}S_{i,3}$. All pair-wise comparisons in group 1 (human vs. primary reservoir, primary reservoir vs. host, and human vs. host) yielded high similarity scores, indicating a high similarity among receptor sequences.

The similarity scores were comparatively low in the non-infection group (g = 2), ranging from 0.092 to 0.440 for ${}^{2}S_{i,1}$, from 0.108 to 0.432 for ${}^{2}S_{i,2}$, and from 0.130 to 0.416 for ${}^{2}S_{i,3}$. For group 2, both the primary host species and non-infected species are listed to illustrate the differences in similarity. In pair-wise comparisons, all the non-infection cases yielded low similarity values, *i.e.*, the receptor sequences differed significantly from each other. We assume that a low similarity in receptor sequences disfavors infection despite the existence of a common receptor. For example, enterovirus infects only Sus scrofa (pig); it does not infectRattus norvegicus (rat) or Homo sapiens (human) because of the high transmission barrier. Similarly, for leukovirus, only Gallus gallus (chicken) is infected as a primary reservoir; because of the high transmission barrier, R. norvegicus and H. sapiens are not infected. These results imply that for non-infection cases, species barriers exist, and the propensity to cross the barrier is determined by the sequence similarity between the potential and primary host receptors. Similarity scores for rabies virus were low between Canis lupus familiaris (domestic dog) and Bos Taurus (domestic cow) (${}^2S_{i,1} = 0.280$, ${}^2S_{i,2} = 0.373$, and ${}^2S_{i,3} = 0.366$) and also between B. taurusand H. sapiens (${}^2S_{i,1} = 0.267$, ${}^2S_{i,2} = 0.371$, and ${}^2S_{i,3} = 0.416$) but were high between C. l. familiaris and H. sapiens (${}^1S_{i,1} = 0.947$, ${}^1S_{i,2} = 0.985$, and ${}^1S_{i,3} = 0.962$). Clearly, C. l. familiaris is the primary reservoir, and transmission of the disease to H. sapiens is possible only because of the high human/dog receptor similarity.

Thus, for particular viruses, transmission of disease may be species-selective, although common receptors exist among species. Furthermore, infection specificity may be determined by the species barrier, which results from receptor differences.

The *x*- and *y*-axes denote ${}^gS_{i,1}$ and ${}^gS_{i,2}$, respectively, where "g" is the group classification. All pair-wise similarity scores are shown. Groups 1, 2 and 3 are each well separated in the colour-coded two-dimensional space. The results provide clear evidence that the receptor sequences from cases of cross-species infection are well separated from those of other infection cases.

From these observations, we conclude that receptor differences are a major contributing factor to the potential of a specific viral strain to cross species barriers for transmission. In other words, the species dependence of infection is indirectly related to the receptor sequence similarity. This finding implies that once the receptor sequences of the primary reservoir and possible hosts are known, we might be able to predict the likelihood of viral disease transmission. The accuracy of these classifications can be judged by subsequent assessment of cases of actual zoonotic transmission to humans.

Three kinds of pair-wise similarity scores (${}^gS_{i,1}$, ${}^gS_{i,2}$, ${}^gS_{i,3}$) are plotted in two dimensional space to show clear differences among groups. Groups 1, 2 and 3 are each well separated; the results show clearly that the receptor sequences from cases of cross-species infection are well distinguished from those of other infection cases. Our analysis revealed significant differences in receptor similarity between infection and non-infection cases. The similarity values, and the experimentally determined group categories were fed into a statistical discriminant analysis to logically predict infection (or zoonosis, in the case of human infection). As described in the Materials and Methods section, the values D_i^2 (i = 1, 2, 3) were calculated from the data to yield results of a specific discriminant analysis.

The statistical discriminant analysis was verified using a test set of four viruses that were deliberately excluded from the training set. The viruses whose groups were predicted using the discriminant analysis. The first virus, feline immunodeficiency virus (FIV), uses Felis catus (domestic cat) as its primary host and CD4 as its receptor. According to the literature, FIV infection of humans is rare but has been reported. Our method categorised this case as near-infection (G = 3). The second virus, classical swine fever virus, is known to be non-zoonotic and was classified as such by our method (G = 2). Thirdly, the encephalomyocarditis virus infects *S.* scrofa but has been known to cause sporadic infections in H. sapiens; it was classified as group 1 (G = 1) by our method. Finally, the Lass virus is known to be zoonotic and was classified as group 1 (G = 1) by our method. The hydrophilic similarity scores (S_1) show less consistency, comparing to the hydrophobic scores (S2), with the predictive values (G). From the result, it could be said that the hydrophobic characteristics of receptor sequence might be the key contributor to the prediction. However, this observation should only be carefully interpreted because the variables (S1, S2, S3) are complementary in the statistical process.

7

Diseases of Cows, Buffaloes and Cattle

The sake of convenience, important diseases of cows, buffaloes and their calves are divided into three categories as follows.

CONTAGIOUS DISEASES

With the introduction of exotic animals in the country, the crossbred cows have increased in number. These crossbred animals are highly susceptible to tropical diseases and climatic variations. Contagious diseases usually take a heavier toll of purebred and crossbred cattle than indigenous. Buffaloes are more resistant to contagious diseases than cows. For example, the disease 'black quarter' is seldom found in buffaloes but takes a heavy toll of milch cows every year. On the other hand, there are a few diseases like haemouhagic secpticaemia which attack both cows and buffaloes but the infection is much heavier on buffaloes. The common contagious diseases that are likely to attack dairy cattle are:

- Bacterial diseases such as anthrax, black quarter, brucelosis, haemorrhagic septicaemia, mastitis and tuberculosis; and
- Viral diseases such as cow pox, foot and mouth disease, rinderpest, etc.

Anthrax

Anthrax is an acute disease having rapidly fatal course. It is the oldest disease known in the cattle. The disease is characterized by septicaemia and sudden death with the exudation of tarry blood from natural orifices of dead animals. This disease has a world-wide distribution. Sporadic cases occur almost throughout India. However, it is more prevalent in certain hot and humid regions of the country. The causative agent of this disease is *Bacillus antkracis*. The organism is relatively large, rod shaped and non-motile. Sporulation occurs outside the body in the presence of oxygen. The spores are highly resistant and are not killed by heat, light and disinfectants. The spores remain viable and infective for several years. The animals get infection by ingestion of food and directly from animal to animal.

Symptoms

- Shivering fits with rise of temperature. Temperature of animal body may go up to 106° F.
- Rumination stops, eyes become red, extremities get cold.
- Breathing is difficult.
- Abdominal pain and tympanities.
- Dung is stained with blood and rectum protrudes.
- Bloody discharge from mouth, nostrils and rectum. The discharge is tarry in colour.
- Staggering gaits, convulsions and animal dies within 24 hours, if the disease is in acute form.

Treatment and Control

The treatment is usually not possible in acute cases. Subacute cases are treated With antibiotics and antianthrax serum. Penicillin and streptomycin in large doses are recommended. Annual vaccination of the animals is recommended in the endemic areas. Alive spore vaccine prepared from a virulent strain of *B. anthracis* is safe for all species. Hygiene is the most important single factor in the prevention of spread of anthrax. As the sporulation of disease causing organisms takes place in the presence of oxygen, the vegetative forms present in the tissues and body fluids will die if the carcass is not opened, but is burnt or buried deep with lime. Destruction of contaminated material and disinfection of equipments and animal shed is also necessary.

Black Quarter

Black quarter is an infectious disease of cattle which may affect even healthy young animals, 6 months to 2 years of age. It is also called black leg disease (since the thigh region of the cattle is called quarter). It is rapidly falal disease caused by a bacterium *Clostridium chauvoei*. It is an anaerobic bacterium which grows only in the absence of oxygen. So bacteria entering the animal body through minute punctured wounds which exclude air get favourable environment to grow. The infection usually takes place through food, water and soil contaminated with black quarter organisms.

Symptoms

- Swelling of the muscular portions usually the quarters, thighs, shoulder, etc.
- Rise in the temperature of animal body and animal becomes dull and goes off-feed.
- Disinclination to move due to swelling. The affected muscle is black in colour.
- Pulse and respiration are accelerated.

- The animal is not able to stand or walk *i.e.* deadly lame.
- It is a Rapidly spreading disease and fatal. The symptoms escape notice because tile animal is found dead all of a sudden.

Treatment and Control

Animals can be protected by suitable vaccination. The vaccine produces immunity in 10-12 days that lasts for about one year. Animals which are showing symptoms should be isolated and treatment done for those showing early symptoms. Penicillin and tetracyclines if given promptly and inoculated into the site of lesions are of value and should be given in normal therapeutic dose. Sulphathiozole and antitoxic sera are also effective. Hygiene and prophylaxis are the methods of control Proper hygiene requires the destruction of carcasses by burning and cleaning and treatment of all wounds.

Haemorrhagic Septicaemia

This disease is widely prevalent in India among cattle and buffaloes. It occurs mostly in acute septicaemic form. Buffaloes are more susceptible to this disease than cattle owing to their greater liking for water and swamps. The disease occurs generally in low lying humid areas and is often seasonal. Outbreak occurs during the periods of highest humidity such as during monsoon. The disease is mainly caused by *Pasteurella multocida*. However, stress due to exhaustion because of excessive work, starvation, chilling or change from rail or road journey may also pre-dispose animals to infection. This accounts for the name 'shipping fever', by which the disease is known in certain countries.

Haemorrhagic septicaemica is carried on through the carriers season after season. The organisms are maintained by carriers during the inter-epidemic period. The infection may occur due to ingestion of polluted grass or water while feeding, grazing and drinking in such places. The organisms get. into the blood through wounds inside the mouth gullet, stomach or intestine and sometimes through wounds in an animal's feet or legs. They may also be conveyed from a diseased animal to a healthy one by bitting insects. The disease may attack throat, lungs or intestines and, as such, three forms of the disease are recognizable namely throat form, lung form and intestinal form.

Symptoms

- Off-feed, high temperature (105 to 108° F) disinclination to move with the herd, shivering, starring coat, and dry hot muzzle.
- Hurried breathing and no milk.
- Dripping of saliva, pain in throat and swelling in the space between jaws, neck, fore limb and brisket.
- The swelling is hot, tense and painful

- One may notice a suffocating cough, difficult breathing and swollen and reddened eyes.
- Colicky(stomach) pain, animal groans and strains when passing faeces which first may be hard and glazed and later on watery, containing blood and mucus shreds. Mortality is very high even 90 per cent.

Treatment and Control

Treatment can be effective if it is carried out in the early stages of the disease but the rapidity of the disease often prevents it. Sulphadimidine and brond spectrum antibiotics are effective. In an endemic area prophylactic vaccination should be carried out annually about a month before the onset of rains. Healthy animals should be isolated from the sick ones and dead animals must be buried six feet deep with lime. Every thing connected with the suffering animals such as ropes, brush, baskets, etc. must be scrubbed, washed With phenyl or even burnt.

BRUCELLOSIS

Brucellosis among cattle and buffaloes produces heavy economic losses due to abortions in behind pregnancy, infertility and reduction in milk production. The disease occurs in most parts of the world. In India, the disease is fairly widely prevelent. The organism responsible for brucellosis or contagious bovine abortion is *Brncella abortus*.

In a herd the disease is introduced by the induction of infected cow or, less commonly, by an infected bull. Under natural condition, the cattle get infection by ingestion of food and water contaminated with uterine discharge of aborted animals. Aborted foetuses, foetal membranes, vaginal discharges, milk and faeces of infected animals contain the organisms, which contaminate the environment, feed and water.

Symptoms

- The cow throws out a dead foetus prematurely but the placenta does not fallout.
- The cow gets sick and there is blood tinged discharge from the vagina.
- Milk yield drops in a lactating cow and the cow goes down in condition.
- There are signs of pregnancy and signs of calving will be noticeable much earlier than the due date and the calf will be thrown out prematurely. In subsequent pregnancies, the foetus is usually carried to full term but second or third abortion may take place in the same cow.
- In the bull, orchitis and epididymitis occur. One or both the scrotal sacs may be affected with acute, painful swelling.

Treatment and Control

The antibiotics are effectual in making the organisms disappear, but they may reappear when the treatment is stopped. Therefore, the treatment is not very successful. Vaccination with *Brucella abortus* strain 19 is helpful in tile control of brucellosis in a herd. The vaccine is inoculated in the calves aged 91-100 days. It stimulates the development of high level of immunity which lasts till fifth pregnancy. The general practices for the control of brucellosis in cattle are hygienic measures, identification and elimination of infected animals from herds. Hygienic measures include isolation of infected animals, disposal of aborted foetuses, placental and uterine discharges and disinfection of contaminated area.

TUBERCULOSIS

Tuberculosis is an antique disease of man, animal and birds. Only a few countries are free of this disease. It is an infectious disease, characterized by slow development of tubercles in almost any organ of the body except the skeletal muscles.

The incidence of tuberculosis depends upon a variety of factors related to husbandry, hygiene and environment. The incidence is high where intensive cattle breeding programmes are carried out in countries where animals are housed indoors during the winter months. Tuberculosis in cattle and buffaloes is caused by the bacterium Tuberculous bacilli. The bacteria enter the body usually through food, sometimes they are directly breathed into the lungs. The diseased animal may contaminate the drinking water and vessels. The excretions like faeces, urine, sputum, etc. of the diseased animal may contain the bacteria and be a source of infection.

Symptoms

The symptoms of tuberculosis depend on the limb affected as any of the organ may be involved. Moreover, an affected animal may show no clinical signs even though it may be severely affected. However, following symptoms may be observed in an affected animal.

- The infected animal gets weaker day by day and becomes inactive and dull.
- Hard, dry cough occasionally but not very pronounced in cows.
- If intestine is affected there will be tympany and diarrhoea.
- If udder is affected, there will be swelling of one or more teats with no pain. The symmetry of the udder may be lost. The milk becomes watery, the quantity reduced and gets curdled on boiling.
- When there is tuberculosis of reproductive system, the abortions occur late in pregnancy. The conception rate is very low. There is thick yellow discharge from the vagina of infected animal.

Treatment and Control

The treatment of tuberculosis is not recommended, as it is very prolonged and the results are unreliable. The animals under treatment are liable to disseminate organisms in milk for human consumption. The tuberculin is widely used throughout the world for the control of tuberculosis in animals. In India where control by test and slaughter is not possible, Bang's method of control has been found to be useful. This method is based on disposing of all the clinical cases. The calves born of tuberculosis-infected cows are free of infection. The animals not showing tile clinical sign are subjected to tuberculin testing. The reactors and non-reactors are kept separately.

The healthy group is tested every 3-6 months. Calves born of healthy animals are allowed to remain with mothers while those from reacting mothers are weaned immediately after birth. This method leads to a progressive increase in the number of healthy animals and decrease of reactors. For preventing the spread of the disease. Hygienic cattle sheds with adequate space ventilation, fresh air, etc. must be provided. The carcasses of affected animals should be burnt or burned, six feet deep with lime.

S. MASTITIS

Mastitis is a disease of udder which usually occurs in heavily milking cows and buffaloes usually after third calving when the milk secretion is at its peak. There is inflamation in the udder due to the infection with micro-organisms. The infection usually occurs due to insanitary conditions of the cattle shed or due to spreading of infection from other infected cows through the hands of the milker. This may occur in anyone quarter of the udder in more than one quarters.

Symptoms

The first symptom is the tendency to kick while being milked in an otherwise quiet animal. Other symptoms are as follows.

- Very severe if the udder gets affected immediately after calving and the teats become swollen.
- Udder is hard and painful with a yellow serum-like fluid to start with, then flocculent material With blood and later on pus.
- There may be slight rise of body temperature. The animal rests her head on her chest or stretches out on the ground.
- Milk is very watery and blood tinged, often curdled, and comes out in clots or shreds, later on as pus.
- The udder loses its normal shape due to swelling of the quarters and displacement of the swollen teats out of position. Milk from the affected udder should not be used for human consumption. But if milk looks clean, healthy and without clots and shreds, it can be boiled and used for calf feeding.

Treatment and Control

The treatment of mastitis is done by antibiotics. Special preparations of antibiotics for infusion into the teats are available. The effectiveness, however, depends on the stage of development of disease. Before infusion of the medicine, the udder is completely milked out. The medicine after introduction into udder is allowed to spread properly by massaging the udder. The control of mastitis depends on sanitary precautions and early detection of affected cases. Small white clots of milk which may appear before the swelling is evidenced can be detected by a 'strip cup'.

This is a wide mouthed cup with a black platform. A black cloth tied over the cup should also be used. Milk from the suspected quarters may be milked directly over the black surface to find out the presence of clots. Another method of testing for mastitis is by using special test papers, which are filter paper strips impregnated with indicators like bromthymol blue. As mastitis milk is slightly aikaline in reaction when a drop of milk is poured over the paper, the colour changes to dark green.

In order to prevent the spread of the disease, the affected cow should be isolated. She should be milked separately by a separate milkman. If a separate milkman is not available, let the available milkman milk healthy cows first and then milk the affected cow last in a separate vessel. Also, keep cattle shed and surrounding clean.

NON-CONTAGIOUS DISEASES

These diseases occur among the animals mostly due to management irregularities, dietetic errors or deficiency in the system or sometimes due to toxic substances. Many of the common non-contagious ailments occur on all kinds of dairy cattle irrespective of breed, age or sex, mostly due to improper care and management.

These non-contagious diseases will not spread from animal to animal nor will they ordinarily prove fatal because most of them are amenable to treatment if promptly detected and treated. The common non-contagious diseases or ailments met with cows and Buffaloes are milk fever, metritis, mammits, tympanities, diarrhoea, constipation, etc.

Ketosis

Ketosis in cattle is a condition occurring due to metabolic disorder. There will be an imbalance between the nutritive intake and nutrition requirement for the body. The condition usually occurs in high producing dairy cattle. There will be low blood glucose level and depletion of glycogen reserves, of the liver. As a result protein is mobilized from the body to be converted into a glucose in the liver. Stored fat of the body is also mobilized. Production of ketone bodies is increased and they accumulate in the blood and urine.

Symptoms

- The affected animal shows decreased appetite.
- Milk yield is suddenly decreased.
- Breath of animal has peculiar sweetish smell.
- The urine also gets a peculiar smell, similar to that of ammonia.
- Body weight of tile animal rapidly reduces.

Treatment and Control

The treatment consists of tile intravenous injection of glucose or fructose. Propionic acid can also be given as sodium propionate. Glycerol or propylene glycol is satisfactory when given orally. To prevent the condition in cows proper care should be taken in the feeding. Feeding balanced ration to high producers and pregnant cows and giving them molasses or jaggery will prevent this condition.

MILK FEVER

Milk fever also known as parturient hypocalcaemia and parturient paresis is a disease which has assumed considerable importance with the development of heavy milking cows. Strictly speaking, the disease is not a fever because the signs of fever such as rise in body temperature are absent. Usually it occurs in cattle which have recently calved, often in the first few days after calving. The disease usually occurs in 5-10 years old cows and is chiefly caused by a sudden decrease in blood calcium level. The volume of milk secretion, particularly when the udder is completely emptied soon after birth may lower the calcium from its normal level of about 10 mg to 3-8 rug per 100 ml of blood. Inorganic phosphorus level is usually decreased from 4 to 6 mg to 1.5-5.0 per 100 ml of blood. The production of colostrum by the animal may also bring about rapid reduction in the concentration of the blood calcium.

Symptoms

- The animal becomes excited and restless, starts shivering with an impression of great discomfort by lashing its tail and paddling its hind feet.
- The animal lies down on her brisket with the head resting on one side of shoulder; snores and moans for a while and later becomes un-conscious.
- Eyes become dull with pupils dilated.
- Breathing becomes deep and slow, pulse is fast but weak, extremities get cold and the temperature falls to 3 or 4 degree below normal.
- Dung and urine are not passed and the gas collects in the stomach.
- The animal loses sense, falls down With partial or complete paralysis of hind quarters and dies within 24 hours if not attended properly.

These symptoms may appear a day prior to calving or even after a month of calving though the majority of cases arise during the first 2 or 3 days of calving.

Treatment and Control

The treatment consists of the sluggish injection during about 20 minutes, into the blood torrent, of 300 ml of 20 per cent solution of calcium borogluconate and the simultaneous injection by the subcutaneous route of 50 to 100 ml of the same solution. The treatment may be repeated in 3 to 4 hours if necessary. In areas of magnesium deficiencies about 11 gm of magnesium sulphate by mouth may be administered to the animal.

The addition of calcium in the ration is contra-indicated as this may even increase the incidence of the disease, particularly if the diet is alkaline. Adding sodium propionate to the feed for 2 or 3 weeks after calving may help to some extent but the availability of sodium propionate is doubtful. The incidence of the disease may be reduced if the cows are only particularly milked during the first three days after parturition and are injected with calcium boro-gluconate immediately after calving.

TYMPANITIS

This is also called bloat or aphara condition. The condition is characterized by the accumulation of gas or foam in the rumen. The abdomen becomes greatly distended especially on the left side in front of the hip bone. The condition arises due to the greedy feeding on lush green especially leguminous fodder, when the animal is in half starved condition. Weak and convalescing animals are more susceptible to this ailment as they cannot digest protein rich feed and the fermentation of undigested feed takes place.

Symptoms

Formation of gas always occurs to a certain extent in the stomach of cattle. But when it occurs in large quantities the following symptoms may be observed.

- Bloated condition of the hollow of the left flank which, if topped with fingers, produces a drum like sound.
- The animal shows the signs of uncareness and pain, frequently lying down and getting up with arched back.
- Rumination is suspended.
- Tension increases leading to breathing difficult when the animal may fall down if neglected and may even die of suffocation.

Treatment and Control

Fermentation can be checked by giving internally anti-fermentative drugs such as turpentine, camphor, asafoetida and ammonium carbonate in right quantities. At the same time, try to remove the gas already formed inside, by cleansing the rectum by a warm soap-water enema. The earliest treatment in a village will be to take 20-30 gm of asafoetida, bum it, powder it and put inside a bread and shove into the oesophagus. Follow it after one hour. In severe cases when the distension is intense, a puncture may be made in the left flank with a trocar and canula to relieve the gas quickly. To prevent this condition, care in feeding lush green fodder such as legumes should be taken. Weak stomach animals should not be given fat and protein rich fodder in large quantities.

Diarrhoea

Diarrhoea is more a symptom of a disease than a disease itself. It is a condition in which dung is passed out in liquid or semi-liquid form, unlike the normal dung which is solid or semisolid. It may be caused by unwholesome feed, spoiled food, toxic food, coarse indigestible fibrous matter or feed adulterated with sand, or stones, polluted water especially after rains which may contain harmful bacteria may also cause diarrhoea. In certain contagious diseases such as rinderpest, worm infestation etc., diarrhoea is a symptom.

Symptoms

- Frequent parrage of watery motions and the soiling of buttocks, thighs and tail.
- The animal is dull, belly tucked up with a staring coat.

Treatment

At first, to remove bacteria and other possible irritants, the animal is given one litre of castor oil emulsion or one litre of linseed oil internally. A few hours later, mix 20 gm catechu, 40 gm chalk powder and 20 gm ginger powder. Feed this mixture with one litre of rice gruel to the animal. This can be repeated twice a day till the diarrhoea stops.

Animals diet should also be corrected by giving dry feed, bran, gram husk, etc. and hay or straw instead of green grass.

CONSTIPATION

Constipation is a condition where the dung is not passed out with ease, in normal quantity of consistency and is usually a symptom seen in all conditions affecting the general health of the animal, sometimes with a rise in body temperature. The causes of constipation are numerous and some important ones are being given below.

- Accumulation of dry and hard undigested feed in the stomach.
- Over-eating of not easily digestible feed such as coarse fibrous roughages.
- Hurried eating-gulping without proper chewing or rumination:
- Inadequate intake of drinking water.

- Increase in body temperature.
- Inefficient functioning of certain internal organs as liver, intestines, etc. Symptoms

Following symptoms are observed in constipation.

- Dung is hard, sometimes even as pellets, passed in small quantities after straining and at times not passed out at all.
- Loss of appetite.
- Animal does not chew, stops rumination and shows signs of pain and discomfort.

Treatment

The affected animal should be given laxative diet like wet wheat bran mash, rice gruel and *canji* water. Supply easily digestible green fodder. Make the animal drink adequate quantities of fresh water, if necessary by adding a little jaggery or salt or both to the water. Evacuate the rectum and give a warm soap water enema. As a first aid relief measure, use following preparation:

0.25 kg of common salt (sodium chloride)
8.25 kg of Epsom salt (magnesium sulphate)
50 gm of powdered ginger
25 gm powdered pepper
10 gm nux-vomica powder.

These ingredients are mixed and divided into two parts. One part is given immediately with feed and the other after a few hours. Six hours later, give half a litre of groundnut oil mixed with 0.25 litre, castor oil, which when given internally, will certainly get the hardened mass out.

To prevent constipation in cattle following precautionary steps may be taken.

- To cows which have a tendency to constipate, supply easily digestible feed stuff With plenty of *conjee* water and drinking water with handful of salt and jaggery.
- Stop feeding straw but give easily digestible succulent grass like paragrass and doobgrass.
- Avoid fibrous feed and add a handful of common salt in the feed or *conjee* water every day as a rule.

EPHEMERAL FEVER

This is a three days sickness condition and is often met with dairy cows especially during winter months. It is also known as rheumatic fever or stiff sickness of cows. This disease is peculiar to cattle and no other animals. The causal organism of this disease is not yet known but it is believed to be of viral origin. Lying on insanitary chill cement floors and exposure to cold may be the predisposing factors.

Symptoms

- Sudden rise of body temperature to 104-1070 F accompanied by muscular pains and lachrymation and salivation.
- The animal goes lame on only one of her legs without any injury or swelling, more commonly the front legs.
- The cow is dull, off feed, stops rumination and breathes hard.
- Milk yield is decreased, swelling is difficult.

Treatment and Control

For treatment of this disease, 2 to 4 ounces of sodium salicylate with equal quantity of Epsom salt and one tea spoonful of ginger with a little jaggery is given by mouth and repeated after six hours. The sick cow is given laxative diet and warm gruel. She is also provided with a comfortable, warm housing. Interavenous injection of sterile sodium salicylate solution by a qualified veterinary doctor, if available, will give immediate relief.

FOOT AND MOUTH DISEASE

The foot and mouth disease is highly communicable disease affecting cloven footed animals. It is characterized by fever, formation of blisters in tile mouth, udder, tests and on the skin between toes and above the hoofs. Animals recovered from the disease present a characteristically rough coat and deformation of the hoof. In India the disease is widespread and assumes a position of importance in livestock industry. The disease is caused by a virus called foot and mouth disease virus. It spreads by direct contact or indirectly through infected water, manure, hay and pastures. It is also conveyed by cattle attendants. Foot and mouth disease occurs in relatively mild form in India and is seldom fatal. It occurs practically all the year round. The disease has been existing for years and, in view of the frequent opportunities for exposure to natural infection, most of the indigenous animals have a great degree of resistance to the disease. Of late, significance is being attached to the airborne transmission of the virus.

Symptoms

The virus gains entry into the blood stream of animals through injury to the lining membranes of, tongue, intestines, clefts of hooves and other similar parts. The incubation period in natural infection is 2.5 days, whereas in artificial infection it is 24 to 48 hours only. After incubation period is complete, following symptoms may be observed.

- Rise of body temperature, dry muzzle, dullness, depression, shivering, staring coat, loss of appetite and stoppage of rumination.
- Slight constipation.
- Dribbling of saliva from tile mouth.

- Formation of blisters on the tongue, gums and cheeks.
- Shaking and kicking of legs and lameness.
- Vesicles at the cleft of the hoof become ulcer like and may get fly blown..
- Vesicles may be seen on the udder and teats also.
- The milk yield comes down in quantity and quality, and the milk coagulates on boiling.
- Tendency of the hooves to get deformed and sometimes the horns of the hooves may be shed. Consequently the animal becomes permanently lame.
- The infected animal cannot be put to hard work, especially in the sun, and it gasps from breath, a condition known as 'panting'. Panting is more severe in the cross-bred animals than in the indigenous ones.

Treatment and Control

No therapeutic agents have been establish till now to cure foot and mouth disease. The use of drugs by field worker is resorted to only as a measure of acidity in the material process of recovery. Thus, the external application of antiseptics contributes to the healing of ulcers and wards off attack by flies. A common and inexpensive dressing for lesions of feet is a mixture of coal tar and copper sulphate in the proportion of 5 : 1. Some other measures to treat the disease are given below.

- Clean the wounds and ulcers in the mouth, udder, teats and feet with 2 per cent potassium permanganate lotion or alum water.
- Decoction of *babool* bark for gargling the mouth and washing the ulcers may also be effective.
- Apply boric acid mixed with glycerin to ulcers in the mouth.
- Foot bath with a disinfectant solution such as cresol or phenol (1 : 100) may be used.
- When maggots are found, custard apple leaves ground into a paste may be applied or a few drops of turpentine are let in.
- Sores on the udder and teats of milch cows should be kept clean and dressed with boric ointment.

JOHNE'S DISEASE

Johne's disease is a serious disease of cattle and buffaloes. It constitutes a potent source of threat to the dairy industry and livestock trade in many countries. It is also known as paratuberculosis as it is produced by Mycobacterium paratuberculosis.. Its insidious nature, protracted and irregular period of incubation, rather vague and indefinite symptoms in the early stage and lack of accurate methods of diagnosis defy the steps to check the indiscriminate movement of the affected animals. The infection bas thus rapidly spread all over the globe. The causative organism is excreted with faeces and will remain alive for a long period. This will cause the spread of disease to other animals.

Symptoms

The disease is very insidious at the onset, and the animal may show clinical symptoms after months or year of infection. In the early stages there is hardly any symptoms which may lead to a positive diagnosis. However, the principal symptoms are given below.

- The affected animal becomes emaciated and bas persistent diarrhoea with bad smell.
- The faeces may often be mixed with flakes of mucous and gas bubbles and are usually passed without straining.
- The appetite is not impaired and there is no fever or pain, the animal has normal body temperature.
- The coat becomes starring and the skin leathery with the progress of the disease.
- Anaemia becomes more marked and on edematous swelling (containing fluid) may develop in the sub-maxillary region.
- The animal goes on loosing condition and at last death occurs due to exhaustion and dehydration.
- In some cases, however, diarrhoea may be absent and dealt may occur due to debility only.

Control and Treatment

At present there appears to be no practical and reliable method of treatment of this disease. The organism is very resistant to chemotherapeutic agents. Because of this the practical utility of treatment in clinical cases is poor. A live vaccine bas been developed. It reduces the incidence of clinical disease. It consists of a non-athogenic strain of Johne's bacillus with an adjutant. The calves soon after birth are inoculated with the vaccine subcutaneously. The vaccinated animals become reactor of John. The vaccination is generally done in heavily infected herd.

General hygiene, test and segregation, slaughter of clinical cases, provision of clean water supply, proper disposal of infected animals improving the general nutritional level, supplementing the rations with minerals, etc. are some of the steps to prevent the spread of the disease. The carcasses of affected animals should always be burnt or buried deep into the ground with lime.

RINDERPEST

Rinderpest is also known as 'cattle plague'. It is the most serious contagious disease of cattle in the country. It is a rapidly fatal disease affecting larger number of annuals and causing enormous losses in infected herds. It affects mainly cattle and other cloven footed animals including buffaloes. Unless special precautions are taken a very large percentage of the affected annuals may die. Rinderpest is caused by an ultra-visible virus.

The virus is present in the blood corpuscles of affected animals. It is excreted through the saliva or discharges from the nose, eyes and the urine and faeces of the affected animals. Outside the body the virus is rapidly destroyed by the sunlight and disinfectants.

The disease spreads through air, contaminated utensils, attendants, etc. It may be introduced in a herd by the inclusion of infected animals.

Symptoms

The incubation period of the disease is 3-7 days. After the incubation period is complete following symptoms are observed.

- High fever, dullness, staring coat, shivering, dry muzzle, and drooping of head and ears.
- Loss of appetite and no rumination.
- Milk secretion in cows stops.
- Salivation, eruptions like bran particles on the inside of lips, gums, dental pad and roof of the mouth.
- Eyelids are swollen, mucus membranes light red and there is lachrymation and nasal discharge.
- Urine becomes dark coloured and scanty.
- First there is constipation which lasts for 2-3 days followed by profuse diarrhoea mixed with mucus and blood with very offensive odour.
- The animal gets exhausted by this time and usually dies after a long period of agony.

Treatment and Control

In earlier days anti-rinderpest serum was used on a large scale for the treatment of the clinical cases of the rinderpest. This practice is now of little value. Anti-rinderpest serum has little or no curative properties once the virus has produced the clinical symptoms in the animal. Hence, the use of this serum is not recommended. Symptomatic treatment with penicillin, streptomycine, sulphadimidine and intestinal antiseptics have no action on tile virus, but may help in tile recovery of tile less severe cases of the rinderpest. It is beneficial to keep the animal on light diet, such as rice gruel with kaolin which act as intestinal astiguents to reduce the intestinal effusions and control diarrhoea. Advanced preventive vaccination if done with freezed dried rinderpest vaccine gives immunity for five years. As there are chances of spreading the disease through Common grazing grounds and at cattle sheds, sufficient precautions should be taken. The affected animals are to be segregated and the newly added animals are to be quarantined for at least 10-15 days.

COW-POX

Cow-pox attacks both cows as well as buffaloes and causes great loss by a rapid fall in the milk yield and by spreading the infection to other animals and human beings. Basically this is a viral disease but secondary infection may be caused by bacteria. The occurrence of cowpox has frequently been associated with the incidence of small-pox in human beings. Instances are on record in which cow-pox in all stages has been transmitted through the milkers. However, conclusive evidence is still lacking as to the existence of any relationship between small pox with outbreaks of cow pox.

Under natural conditions the infection takes place through inoculation by the cutaneaus route and readily spreads from one animal to another through the agency of milkers. The occurrence of this disease is not often reported since it is localised only to the teats and udder, and occasionally to the hairless parts of the body. It does not assume a serious form unless the lesions are infected with pyogenic (pus producing) organisms.

Symptoms

- The first symptom is the tendency to kick while being milked, by an otherwise quiet cow.
- Teats get swollen, hard and painful.
- There is rise of body temperature, loss of appetite and stopping of rumination.
- Eruptions appear on the skin of the udder and teats developing into vesicles, pustules and scabs by stages, sometimes on the thighs and abdomen,
- Reduction in milk yield.

Treatment and Control

The lesions heal by themselves in the normal course and the adoption of special measures is not generally required, only the usual rules of hygien need to be observed. The lesions should be cleaned with a 1 : 1000 solution of potassium permanganate followed by the application of an antiseptic ointment such as 1 : 10 boric acid.

The affected animals should be isolated and milked by separate milkers. Milk from affected animals should be boiled before use. If the disease assumes serious proportions, vaccination may be undertaken.

COMMON DISEASES/AILMENTS OF SUCKLING CALVES

Like grown up cattle, suckling calves also get infected with a number of diseases or ailments of various nature. Some of them such as umbilical abscess, blindness, white scour, dysentery, etc. may prove quite disastrous. These diseases are briefly described here.

Umbilical Abscess

This condition is seen in young suckling calves as on infected abscess at or near the umbilical opening. This may be caused due to neglect by the owner in not keeping the calf under hygienic condition or providing any bedding or by leaving the cut end of the umbilicus exposed to flies and bacteria Without any antiseptic dressing. In this condition following symptoms are observed.

- The part is hard, hot and painful with a foul smelling discharge at the opening.
- Slight rise in body temperature.
- Sometimes, swelling and pain at the leg joint may also occur. For the treatment, following steps should be taken.

Clean tile part with a mild disinfectant such as a few drops of tincture of iodine or phenyl in clean water. Foment and dress it with any antiseptic powder like Mycodenn or Nebasulph. Application of Horexane and Himax ointment will also be useful.

Keep the stall floor clean with disinfecting lotion like Phenyl, Salvon or ettol and provide a layer of dry straw or leaves as bedding for the calves.

Pneumonia

Pneumonia occurs in young calves soon after birth or a few days later, especially during winter season. It is caused due to exposure to cold weather or by bacterial infection. It may also occur due to worm infestation in the lungs. Following symptoms are observed.

- Rise in body temperature.
- Thick nasal discharge.
- Laboured respiration.
- Indigestion and diarrhoea.

To treat pneumonia, following measures are suggested.

- Steron inhalation with, eucalyptus or terpentine.
- Dry, warm and well ventilated stall with dry bedding.
- Antibiotic injection such as Penicillin or Amphicilline.
- The sick calf should be given warm, nutritious and easily digestible diet.

Dysentery

Dysentery in calves is caused due to irritation in intestine or sometimes due to wonns. It may also be caused by Coccidia, a protozoan parasite.

- The calf strains, it does not eat and has abdominal pain.
- Small quantities of dung are passed with mucus and blood in it. Treatment:
  1. Give castor oil emulsion (2 to 4 ounces).
  2. Give sulphaguanidine (2-4 tablets) twice a day for three days.

3. Feed 3-4 sour pomegranate fruits, chopped with the rind, daily for three or four days.
4. Feed the calf with rice gruel to which 10 gm of fenugreek has been added and boiled.

White Scour (Diarrhoea)

White scour disease is fairly common in most calves, especially in buffalo calves. It occurs usually within the first few weeks of birth. The condition always indicates bad management. Following factors may be responsible for this condition.

- Over drinking of mother's milk, especially when it is rich in butter fat.
- Improper feeding *i.e.* frequent feeding, more in quantity, milk rich in fat, cold and polluted milk.
- Lying in mud floor especially when it is wet and insanitary.
- Heavy worm infestation in the intestines.
- Ark, dirty and ill-ventilated calfshed.

The main symptoms of this disease are severe-whitish diarrhoea, dullness, progressive emaciation (weakness), etc. The dung of the calf is loose and watery. Often it is whitish or clayish in colour with an offensive odour. The calf strains and moarns with pain. For the treatment of white scour in calves, following steps are taken.

- Housing on dry hygienic floors, preferably with some bedding of dry straw or dry leaves.
- Good ventilation and sunlight are essential.
- Wean the sick calf, withhold whole milk and give only skimmed milk or gruel made of ragi flour and broken rice.
- Give catechu, one and half tea spoonful with equal quantity of chalk powder or jaggerry.
- Administer orally the tablets of sulphromezathine, two tablets twice a day for three days
- Add two teaspoonfuls of Aurofac, Neften or lime water to the feed for a month.
- Use a mouth gag to the calf at non-feeding time to prevent it from slicking the walls and floors.

Blindness

Blindness is a condition commonly seen in new born calves. It is caused due to lack of vitamin A. The calf develops following symptoms.

Symptoms

- The calf appears normal in all respects but standing on its feet in about half an hour as any other normal calf, it is not able to find the udder for want of vision and so keeps knocking about.

- The eyes look normal but when an attempt is made to touch the eyes with fingers or any other object, the eyelids will not close.

Treatment

To treat blindness in calves, following measures are suggested.

- Give one or two teaspoonful of good quality shark liver oil in feed at least twice a day for a month or more.
- Vitamin A preparations which are easily available in the market can be added to the feed for a month or more.
- Feeds rich in vitamin A such as yellow maize meal, chopped carrots, etc. should be fed till weaning.
- A few injections of vitamin A preparations like Prepaline, should be administered to care the calf completely.

Worm Infestation

Worm infestation is a common problem in calves, sometimes even at birth. It occurs almost invariably in all buffalo calves. It is caused due to parasitic infestation, mostly roundworms and occasionally tapeworms, hookworms and even liverflukes. Secondary factors such as insanitary conditions of housing and ill management may also be responsible. In villages, it may also be I caused due to polluted water and fodder from marshy areas. Following symptoms may be observed.

- The infested calves get diarrhoea.
- The calf gets rough coat and depressed appetite.
- The affected calf develops the tendency to lick mud and eat any thing.
- The abdomen of affected calf becomes pendulous. Grinding of teeth may also be observed.
- Sometimes even worms may be passed out with dung.

To get rid of this affliction, the dung of affected calves should be got examined. Dewomung the calf with known anthelmintics (worm dose) such as tablets of Decaris, Helmonil, oil of turpentine arecanut, asafoetida or tender leaves of neem with salt should be practiced.

Skin Disease (Ring Worm)

Skin disease is a common occurrence among young suckling calves. Housing calves under unhygienic conditions, improper grooming and brushing and parasites like ticks, flies, lice, etc. are the factors responsible for this condition. In affected calves, hairless patches on the head, neck, face and sometimes on the legs are formed. They also get itching and irritation, which can be noticed. The animal rubs against walls, posts and trees. Affected animals should be isolated. Clip the hairs on the affected parts, remove crusts, clean the surface and paint it with tincture iodine. The calves should be given good nourishing diet and cod liver oil. Apply antifungal ointments such as Mycozol on the affected portion. Sometimes painting the affected portion with the juice of *Leucas aspera* or applying oil obtained by burning coconut oil, are also effective. Livestock has played a very important role in the Indian economy ever since civilisation. With the increasing human population in India, demand for animal protein is increasing, due to social and economic reasons. Milk and milk products are important constituents of human diet. Some of the population also take animal meat, especially of buffaloes, due to low cost and taste. Thus control and eradication of diseases of cattle and buffaloes are very important considerations. Now with the introduction of some of the foreign breeds of cattle and their crosses, which are highly susceptible to diseases, tile demand for efficient health cover has increased considerably in order to maintain them in good health and production. For tins, control and eradication programme for these cattle diseases has become very important considering the economies of cattle and buffalo industry.

Prevention, control and eradication are the three basic methods used in dealing with any disease in the cattle population. These three methods are applied depending upon the economic importance of disease and investment cost available for the control programme. For prevention, control and eradication of all contagious diseases (which are mainly caused by virus or bacteria), a potent vaccine is very essential. For most of these contagious diseases, vaccines are available. In susceptible areas or herds, vaccination programme should be carried out well in advance.

It is also imperative that cattle and buffalo owner should know all the signs of good health and understand the condition of his animals especially when they manifest symptoms of diseased condition, so that he can take effective steps in time to get his animal treated or to control the spread of infection. Every time it may not be possible for the owner to attend his sick animal and as such, he should take the help of a qualified veterinary doctor in proper time. However, there are certain diseases for which there is no any vaccine or scientific treatment course. In such a case, improvement in hygienic conditions, management practices and feeding standards is very important.

DISEASE PREVENTION AND CONTROL FOR CATTLE

Monitoring your livestock and following good farming practices are the best ways to reduce the risk of disease among your cattle. You can find out more about measures to prevent disease on your farm in our section on controlling disease.

Buying, Registering, Identifying and moving Cattle

There are rules which you must follow for registering cattle, and when moving them. These procedures make it easier to trace and identify infected animals in the event of a disease outbreak.

When restocking your herd, you should:

- Consult your vet to help you develop a plan for evaluating prospective purchases
- Buy health scheme animals-whenever possible-that have been certified as free of specific diseases
- Recognise the risks in buying older animals.

Biosecurity and Hygiene

Good biosecurity is a vital part of keeping disease away from your animals. This will also protect the health of your workers and any members of the public who may visit your farm.

Your general biosecurity measures should include:

- Restricting and controlling movements of people, vehicles and equipment into areas where your cattle are kept
- Cleaning and disinfecting equipment, vehicles, protective clothing and footwear before and after contact with farm animals.

The presence of a disease may not always be apparent-particularly in the early stages-so the measures above need to be part of your routine.

Health and Welfare Programmes

You should seek veterinary and technical advice to create a written health and welfare programme for your cattle. This should be reviewed and updated annually. As a minimum, it should include your:

- Vaccination policy and timing
- Parasite control procedures-internal and external.

Drugs are useful for preventing disease in cattle, but you must use them responsibly and record all usage. You can reduce the risk of disease and improve the performance of your livestock by using farm health planning techniques.

One good way to improve health planning is to benchmark your cattle's health and business practices against other local farmers.

Veterinary Surveillance

As part of its disease prevention strategy, Defra constantly collects information about incidences of disease in animals. This monitoring enables any important or unusual outbreaks to be detected quickly-so that appropriate action can be taken.

The Single Payment Scheme (SPS) and Farmed Animal Requirements

If you receive SPS or other direct payments, you must comply with Statutory Management Requirements (SMRs). SMR 7, for example, sets out the requirements for cattle identification and registration whereas SMRs 10-15 are about disease prevention through the feeding and medicinal treatment restrictions set out in this guide.

DISPOSAL OF FALLEN STOCK

'Fallen stock' is any animal that has died of natural causes or disease on a farm or that has been killed on a farm for reasons other than for human consumption. To bury or burn fallen stock on farm due to the risk of spreading disease through residues in the soil, groundwater or air pollution. This ban also covers animal by-products (ABPs), including afterbirth and stillborn animals.

The only exceptions to this ban are:

- In remote areas-parts of the Highlands and Islands of Scotland, Bardsey Island and Caldy Island in Wales, and the Scilly Isles, and Lundy Island
- During outbreaks of notifiable disease-if there is a lack of capacity at rendering plants and incinerators, or if transporting carcasses would spread disease.

All fallen farm animals-including stillborn animals-must be collected by an approved transporter and taken for disposal or treatment to an approved:

- Knacker
- Hunt kennel
- Maggot farm
- Incinerator
- Renderer.

Fallen stock must be collected, identified and transported without 'undue delay'. This means as soon as is reasonably practical under the circumstances (usually within 48hrs of death). ABPs must be transported in covered leak-proof containers/vehicles and be accompanied by a commercial document.

National Fallen Stock Scheme (NFSCo)

You can dispose of fallen stock privately at an approved destination, or you can make arrangements through the NFSCo. As a farmer, you can register with the scheme for the collection and disposal of your fallen stock. Members receive a list of the approved collectors operating in their area and the prices that they charge. Members can contact a registered collector of their choice whenever they have fallen stock. You can find information on fallen stock on the National Fallen Stock Scheme website-Opens in a new window.

ANIMAL HEALTH OFFICE (AHO)

If you want to make your own arrangements, your local AHO will provide a list of approved knackers, hunt kennels, maggot farms, incinerators or renderers. If you want to burn animal carcasses in your own on-farm incinerator, your incinerator must comply with ABP controls and environmental permitting requirements. It must also be approved by the AHO. ABPs can be incinerated on the page on Compositing and incinerating animal by-products in our guide on dealing with animal by-products.

Rules for Bovine Spongiform Encephalopathy (BSE) Testing Fallen Cattle

BSE is part of a family of diseases known as Transmissible Spongiform Encephalopathies (TSEs). These are fatal brain diseases, which include classical and atypical scrapie in sheep and goats, and BSE in cattle. TSEs in fallen cattle to assess the level of these diseases in your animals. This will enable you to assess how effective disease control measures would be and help you to ensure any controls are proportionate to the risk.

Fallen stock must be tested for BSE if they are:

- Over 48 months of age
- Over 24 months of age and born outside the European Union.

Owners in the UK must contact a collector within 24 hours of their animal's death, to arrange delivery to an approved sampling site within 72 hours of the animal's death. If you are delivering the carcass yourself, you must contact an approved sampling site to agree this within 24 hours and you must deliver the carcass within 72 hours of the animal's death.

Fallen cattle aged 48 months or under should be disposed of in accordance with ABP legislation. Animal carcasses or parts of animal carcasses suspected or confirmed as infected by a TSE are category 1 ABPs-the highest risk category. They must be disposed of by incineration, or processing (rendering) followed by incineration. You can find out about destroying TSE-infected carcasses and specified risk material in our guide on keeping livestock healthy-disease controls, prevention, notification and restrictions.

DISEASES OF CATTLES EYE

INJURIES OF EYES

Cattle are liable to receive injuries to the eyes from brush and weeds while grazing; also from chaff and other substances getting into the eye, causing it to become inflamed and throw off a watery discharge.

Treatment

Examine the eye carefully to remove any foreign body that may be there. If there is much pain, bathe the eye for ten or fifteen minutes night and morning with water as hot as can be borne with the hand. If the irritation is only slight, cold-water bathing will be all that is necessary. In very severe cases the animal should be kept in a dark stable in the brightest part of the day.

BLEEDING FUNGUS OF EYES

This is a cancerous growth which sometimes affects the eyes of cattle. A fungus growth grows in the eye and tissues around it, protruding from the orbital fossa and sometimes crowding the eye out with it. There is a constant discharge and a fetid odour from it.

Treatment

The only remedy is to remove the entire growth, eye and all tissues involved; scrape the bone clean and then apply powdered sulphate of copper to all raw surface and fill the cavity with cotton. In twenty-four hours the cotton should be taken out and the wound washed and dressed once a day thereafter with the following ointment: Pine tar and lard, of each, four ounces; acetate of copper, half an ounce; mix.

CONTAGIOUS OPHTHALMIA IN CATTLE

Contagious ophthalmia in cattle is comparatively a new disease in Kansas, only having made its appearance about five or six years ago. It does not sweep the entire country, but is rather enzootic in its habits, although when it enters a herd it continues until all have had it. It is an inflammatory condition involving some of the inner structures of the eye and the eyelids.

Symptoms

There will be a watery discharge from the eye; the lids will be kept closed. As the disease progresses, the eyelids will become swollen; the cornea will have a whitish colour, and the animal will grow dumpish and lose its appetite. In very severe cases the swelling continues until the eye-ball is ruptured and its contents discharged, precluding all possibility of restoring sight.

Treatment

Place the patient in a dark, cool stable; four ounces of Epsom salt should be dissolved in a quart of water and given as a drench twice a day; if the bowels become too loose, give only once a day. In mild cases, bathing twice a day with cold water may reduce the inflammation; but, if there is much swelling, bathe twice a day with hot water, and each time wipe the skin dry and apply a little of the following, all around and in the eye: Nitrate of potash, forty grains; sulphate of zinc, forty grains; fluid extract of belladonna, four drachms; water, one pint. If the eye-ball ruptures, syringe it out with the same lotion.

PARASITIC DISEASES OF CATTLE

ACTINOMYCOSIS-LUMPY-JAW

Actinomycosis, or lumpy-jaw, is characterized by swellings on the jaws. The swelling may be on either upper or lower jaw and may be somewhat soft at first, but soon grows hard, as the disease affects the bone. It is due to the parasite actinomysis, a vegetable fungus supposed to be on the fodder or other food and enters the animal organism through some abrasion of the skin. It sometimes affects the tongue, causing it to thicken up and become hard, in which condition it is the so-called 'woody-tongue.' The general health of the animal does not appear to be much affected by the disease as long as the animal can masticate its food, but after a time the jaw becomes affected to such an extent that the animal can scarcely eat, then it soon becomes emaciated. Theorists differ as to whether the flesh of such an animal is fit for food. The safest plan is to let it alone.

Treatment

There have been many methods of treatment prescribed and some 'sure cures'(?) advertised. But the only thing most of them affect, with any certainty, is the owner's pocket book. The best treatment known for this disease at the present time is that recommended by the Bureau of Animal Industry, which is as follows: Give a daily dose of iodide of potassium, allowing fifteen grains for each one hundred pounds of the animal's weight. This may be dissolved in water and given as a drench, and should be continued until symptoms of iodism are produced, which will be in from ten to fourteen days.

The animal will become languid and disinclined to move about; the appetite will fail and there will be a discharge from the eyes and nostrils. In some subjects, not all, there will be a vesicular eruption of the skin; there will be abstinence from water, and elevation of temperature. When these symptoms occur, the iodide should be withheld for a few days until the appetite returns and the other symptoms subside, when the treatment should be repeated as before. The treatment should be continued in this manner until the enlargement begins to decrease in size, which may be two or three months.

Tuberculosis Consumption

Tuberculosis, or pulmonary consumption, is an infectious disease due to the bacillus tuberculosis which, through various channels, invades the animal's body. The most important consideration, with regard to animal tuberculosis, is the bearing that it has upon the health of the human family. That animal and human tuberculosis, or consumption, are identically the same has become an established fact; and it is also the belief of the most scientific investigators that animal, and especially bovine, tuberculosis is, to an enormous extent, responsible for the same disease in the human race.

The disease may be communicated from the animal to the human being in various ways. The milk of diseased animals is known to be one of the most prolific sources of communication. The flesh, though not always, has been known to contain the bacilli, and, as many people eat their meat 'rare done,' it thus becomes a source of danger. The discharge from the nostrils and from abscesses of diseased animals falling upon the ground or upon stable floors, becoming dry and being taken up as fine dust by the moving air, may be inhaled by other cattle and also by human beings. Milk, containing the bacilli, spilled upon floors of dwellings and dairy rooms, in this manner becomes a source of communication by inhalation, even to those who do not drink it.

A most serious aspect of this subject is the fact that this disease is not easily detected in its first stages; its progress is often slow at first, the animal being apparently in fair health for a year or two after the disease has begun its work. A case which came under my own observation was that of a cow belonging to Dr. S. D. Ross of Manhattan, Kansas

The subject was a high-grade Jersey cow, kept to supply the family with milk. Some time during the spring of 1892 the writer was called in and found the cow slightly indisposed, as if from a cold. There was affected breathing, and a slight wheezing sound in the lungs. Tuberculosis was mentioned at the time, but as we had no microscope and did not then know the use of tuberculin, no test was made; and, as after a few doses of stimulants and tonics, the cow seemed to recover, nothing more was thought of the case until two years after, when your humble servant was again called to deliver a calf from the same cow.

The cow was very thin in flesh, and Dr. Ross stated that she had not been doing well for a month or two. The calf was dead, but was soon delivered, and when the placenta was taken away its uterine surface was found thickly studded with nodules and small ulcers. Having access this time to a microscope, the writer prepared several specimens, in each of which were countless numbers of the bacilli of tuberculosis.

Symptoms

The symptoms are often so slight at first as to be almost unnoticed, or may be mistaken for a common cold. The animal may be slightly off its feed for a few days; the muzzle may be dry, and, if a cow, there may be a decrease in the quantity of milk. Good nursing, with a few doses of medicine, soon checks the disease, and the animal is all right for a while. These attacks become more frequent and each one leaves the system in a worse condition until finally the symptoms never abate entirely.

The hair will begin to look rough; the appetite will be capricious; the breathing will be affected; the animal cannot stand exertion; the ear placed against the side of the chest will detect a wheezing or whistling sound; there may be a cough and, as the disease progresses, there may be a discharge from the nostrils with a fetid odour; as the symptoms become more aggravated, there will be great emaciation; there will be complication with bowel troubles, and some times tubercular tumors or abscesses form about the head and neck and other parts of the body.

But, though these symptoms are sufficient to excite suspicion, the only true test is with the microscope, or by the hypodermic injection of tuberculin, both of which methods require experience. The disease is incurable, and as soon as its existence is proven, the affected animal should be destroyed and the carcass burned. The milk of tuberculous cows should not even be fed to pigs. The writer has seen several well-marked cases of the disease in swine.

Eruptive Aptha, or Pseudo-Foot-and-Mouth Disease

This disease first made its appearance in the State of Missouri a few years ago and was thoroughly investigated by Dr. Paul Faquin, State Veterinarian of Missouri at that time, and found to be due to some vegetable parasite. Through a misconstruction of some statement made by the doctor, a report became widely circulated that the cases were of the true foot-and-mouth disease, and considerable excitement prevailed for some time; but this false report was corrected by the Secretary of Agriculture at Washington, D. C., and the excitement, as well as the disease, soon became quiet.

Nearly every season since, there have been a few cases in Missouri and Kansas (and probably other Western States), but it has never become, serious, and if any animal ever died with the disease it must have been because the owner neglected to furnish proper food for it while its mouth was sore, and it starved to death.

The disease is simply an eruption of the mucous membrane of the mouth and of the skin of the body; and although the two diseases have many symptoms in common, yet it lacks the virulence of the foot-and-mouth disease of European countries.

Symptoms

Small blisters form on the tongue and inside of the lips; these burst open, forming small sores, and sometimes the lips become dry and cracked. The animal eats readily if food is placed well back on the tongue, but cannot take it up because of the soreness of the mouth. The jaws are kept moving and the saliva flows freely. Small eruptions appear over the body; and about the udder and other parts where the hair is thin, the skin becomes of a reddish colour. The soreness on the skin extends down to the feet and sometimes cracks form around the hoofs, and, in severe cases, the hoofs come off. The animal often gets very lame and is inclined to lie down a great deal.

Treatment

If there is diarrhoea, give a dose of raw oil; if no diarrhoea exists, give a dose of Epsom salt sufficient to open the bowels moderately; give hyposulphite of soda in the drinking water to the amount of three ounces in twenty-four hours. Swab the mouth three times a day with the following: Alum, one ounce; water, one quart; mix.

Sponge all sores on the body and feet with a solution of sulphate of copper, one ounce, to water, one quart. Keep the patient in a cool, shady place, and feed on gruel, bran mash, boiled oats and other soft food.

CATTLE HEALTH

Figure 1.1 : Cattle

As a cattle farmer, it's essential take the necessary steps to protect the health of animals. Maintaining good health is important to ensure acceptable standards of animal welfare, but such measures will also maximise the productivity.

This conduct has information about the main diseases affecting cattle, including notifiable diseases, which must be reported, and zoonoses, which can infect humans. It covers general disease prevention and legal controls to prevent specific risks, as well as describing your legal responsibilities relating to cattle feed, medicines and hormonal treatments.

CATTLE HEALTH AND CROSS COMPLIANCE

To succeed for full payment under the Single Payment Scheme and other direct payments-*e.g.* the Environmental Stewardship schemes-you must meet all relevant cross compliance requirements. These requirements are split into two types:

- Statutory Management Requirements (SMRs)
- Requirements to keep your land in Good Agricultural and Environmental Condition (GAECs).

The aim of these requirements is to stop the illegal use of substances that have a hormonal or thyrostatic action and beta-agonists in stock farming. They also prevent residues that these substances leave in meat-and other foodstuffs-from entering the human or animal food chain. The aim of SMRs 12, 13, 14 and 15 is to minimise the risk posed to human and animal health by certain transmissible spongiform encephalopathies (TSEs), foot and mouth, bluetongue and other animal diseases. These will apply to you if you keep cattle.

CRITICAL ILLNESSES OF CATTLE

Looking after animals properly and monitoring them regularly for signs of illness are the best ways of preventing disease, and of controlling its spread if there is an outbreak. The suspect signs of a notifiable disease, you must immediately notify the duty vet at your local Animal Health Office (AHO).

Cattle are vulnerable to many endemic and exotic diseases, some of which might be notified to Animal Health if suspected:

- Endemic diseases are diseases which commonly exist within Great Britain, *e.g.* common diseases such as Bovine Viral Diarrhoea (BVD), Infectious Bovine Rhinotracheitis (IBR) and Johne's disease. They can be both notifiable-those that the law requires to be reported to the veterinary authorities-and non-notifiable.
- Notifiable diseases can be both endemic, such as tuberculosis, and exotic, meaning that they are not normally found in Great Britain, such as foot and mouth and bluetongue. Outbreaks of these diseases are subject to national control policies and international trade rules. They must be reported immediately to the duty vet at your local AHO.

Endemic Diseases

The most important endemic diseases to check for in cattle are:

- BVD and BVD2-bovine viral diarrhoea
- IBR-infectious bovine rhinotracheitis
- Johne's disease
- Certain forms of tuberculosis
- Salmonella
- Mastitis
- Other viral, bacterial and mycoplasma pneumonia.

Notifiable Diseases

The main notifiable diseases which affect cattle are:

- Anthrax
- Aujesky's disease
- Bluetongue
- BSE (bovine spongiform encephalopathy)
- Brucellosis (brucella abortus)
- Contagious bovine pleuro-pneumonia
- EBL (enzootic bovine leukosis)
- Foot and mouth disease
- Lumpy-skin disease
- Pleuropneumonia
- Rabies
- Rift-valley fever
- Rinderpest
- Bovine tuberculosis
- Vesicular stomatitis
- Warble fly.

PREVENTION AND CONTROL OF CATTLE

Cattle farmers must take steps to minimise the risk posed to human and animal health by brucellosis and transmissible spongiform encephalopathies

(TSEs), such as BSE (bovine spongiform encephalopathy) and its human variant.

PREVENTION AND CONTROL OF BRUCELLOSIS

Brucellosis of cattle, known as contagious abortion, is caused by infection with the bacterium 'Brucella abortus', which can also infect people. Cases have been recorded in this country recently-despite its eradication from the UK-through the movement of infected cattle from other countries. You should therefore work out proper biosecurity with your vet before you bring imported cattle on-farm. The disease causes abortion-or early calving-of recently infected animals, with large amounts of infectious material produced. The bacterium is also excreted in milk. People can become infected by drinking untreated milk and by contact with infected animals. The bacterium is destroyed when milk is pasteurised.

All abortions in cattle, which occur at a certain stage of pregnancy, should be reported to Animal Health. The law requires cattle keepers to report every abortion or premature calving to Animal Health. An abortion or premature calving takes place less than 271 days after service or 265 days after implantation-or transfer of an embryo-whether the calf is born dead or alive. Although blood tests for brucellosis are no longer carried out routinely on beef cattle, the following surveillance and risk management measures continue to apply:

- Import controls and certification
- Exporting country alerts of brucellosis breakdowns in exporting herds
- Post-import check testing
- Post-calving check testing
- Abortion reporting and investigations.

Any suspicion of this notifiable disease, must contact Animal Health Office (AHO) immediately. Treatment for brucellosis of cattle is not permitted-all infected cattle as well as cattle that have been exposed to infection must be slaughtered. Animal Health Officers have the right to serve a notice on the owner or person in charge of the affected bovine animals to ban their movement from their current premises or require their movement to other premises. Where a notice applies, no manure, slurry or other animal waste can be removed from those premises except under a specific licence.

Prevention and Control of TSE

To reduce the risk of TSEs, you must:

- Not feed animal protein or any feeds containing animal protein to cattle, sheep, goats, camelids, bison, buffalo, deer, antelope and wildebeest-except for milk, milk-based and colostrums, eggs and egg products, gelatine from non-ruminants, hydrolysed proteins derived from non-ruminants or from ruminant hides and skins, and liquid milk replacers for unweaned ruminants containing fishmeal, if registered by Defra.

- Not use feed products containing prohibited proteins or mix prohibited proteins with feeding stuffs such as bonemeal or poultry meal
- Not use restricted proteins to produce feed for non-ruminants such as pigs and poultry-unless you have received authorisation from Defra
- Not use feed products containing restricted proteins on a farm where there are ruminant species present-unless you have received registration from Defra
- Notify the duty vet at your local AHO immediately if you know or suspect that an animal or carcass in your possession-or under your charge-is infected with a TSE
- Fully comply with any movement restrictions
- Fully comply with any order to slaughter and destroy any animal
- Fully comply with any other notices served by an inspector.

There are specific rules to prevent the spread of BSE and other TSEs through the use of animal by-products in animal feed.

PREVENTION AND CONTROL OF TUBERCULOSIS IN CATTLE

Bovine tuberculosis (bTB) is a pressing animal health issue. The incidence rate of bTB in cattle in England and Wales has been rising for 25 years and has worsened since the 2001 Foot and Mouth Disease outbreak. The disease is a largely regional problem, concentrated in England in the South West and West Midlands. As the disease can spread to humans through contaminated milk and dairy products it is unlawful for milk-raw or pasteurised-from any animal testing positive for bTB to be used for human consumption. Milk from the rest of the herd can still be sold for human consumption-as long as it's heat-treated.

Testing

Cattle with bTB are most often identified through testing using the tuberculin skin test before they develop clinical signs. This is because the disease usually progresses slowly and it can take some time for clinical signs to appear. Clinical signs of advanced bTB include:

- Weakness
- Emaciation
- Difficulty breathing
- Enlarged lymph nodes
- Coughing.

The veterinary manager of your local Animal Health Office (AHO) immediately if you know or suspect that an animal or carcass in your possession-or under your charge-is infected with bTB.

All cattle herds in Great Britain-except some beef fattening units-are regularly tested for bTB, as part of four different testing programmes:

- Compulsory regular testing using the tuberculin skin test-carried out at the Government's expense every one to four years, with the frequency determined by how widespread bTB is in a particular region
- Gamma interferon blood testing – in the circumstances outlined below.
- High health risk testing-additional annual tests where there is a potentially higher risk to public or animal health, *e.g.* producers of raw drinking milk, unpasteurised dairy products, dealers' herds, bull hirers and regular buyers of Irish cattle
- Pre-movement testing-compulsory tests before cattle aged 42 days or more can be moved from a high risk area
- Post-mortem testing-inspection carried out by the Meat Hygiene Service on all cattle at slaughterhouses, which enables bTB cases to be traced back to the herd of origin.

Tuberculin Skin Test

The tuberculin skin test is the primary screening method for bTB in Great Britain. This test is used throughout the world to screen cattle, other animals and people for TB. Testing is carried out by government-approved testers and supervised by government-approved vets. Each test has three possible outcomes:

- Positive-the reactor animal is isolated from the rest of the herd and slaughtered
- Inconclusive-the animal is re-tested
- Negative.

Gamma Interferon Blood Test

The supplementary sensitive gamma interferon (g-IFN) diagnostic blood test is used in prescribed circumstances alongside the tuberculin skin test, to improve the sensitivity of the testing regime and identify more infected animals more quickly.

Using both tests in this way can help to speed up the resolution of confirmed TB breakdowns by identifying as many infected cattle as possible at the earliest opportunity. The use of g-IFN test is mandatory in the following prescribed circumstances:

- All confirmed new incidents (CNI) in 3 or 4 yearly tested herds, including those that fail to resolve through repeated skin testing or where complete or partial de-population is contemplated
- Confirmed TB incidents that have failed to resolve through repeated skin testing, in 1 and 2 yearly tested herds, including those herds where complete or partial de-population is contemplated.

Additionally, the test is used occasionally to enhance specificity in the following limited circumstances:

- Non-specific reactor procedure for unconfirmed breakdowns in 2, 3, or 4 yearly tested herds
- Suspected fraudulent reactors

Pre-Movement Testing

In addition to the above, Pre-Movement testing is a statutory requirement: cattle 42 days old and over moving from a 1 or 2 yearly tested herd must have tested negative to a bTB test within 60 days prior to movement.

Cattle Compensation

Government compensation is paid to owners of cattle compulsorily slaughtered for bTB control purposes. Compensation in England is determined primarily using table values, which reflect the average sales price of bovine animals in 47 different categories. The categories are based on the animal's age, gender, type (dairy or beef) and status (pedigree or non-pedigree).

Vaccination

Vaccination of either cattle or wildlife is a potential long-term option for reducing the risk of bovine TB in Great Britain. However, vaccines can never represent a single answer to the problem of bovine TB. Vaccination is a risk reduction measure, most likely to be successful in controlling bovine TB when used alongside other disease control. The first injectable badger vaccine was licensed in March 2010 and is available for use on prescription. Research is continuing into producing a licensed cattle vaccine with differential test and an oral badger vaccine.

Badgers and bTB

Allow controlled culling to be carried out by groups of farmers and landowners, as part of a science-led and carefully managed policy of badger control. Licences will be issued by Natural England under the Protection of Badgers Act 1992 to enable groups of farmers and landowners in the worst affected areas, to reduce badger populations for the purpose of preventing the spread of disease at their own expense. Culling will be piloted of the policy in two areas in 2012 to monitor the effectiveness, humaneness and safety of controlled shooting. Subject to Natural England's decisions on licence applications, the two areas will be West Gloucestershire and West Somerset. An independent panel of experts has been appointed to report to ministers on the findings of the monitoring of the badger culling pilots by the end of 2012. If Ministers decide to proceed following the pilots, a maximum of ten licences will be granted to start each year. A five-year Badger Vaccine

Deployment Project (BVDP) is taking place in one area of Gloucestershire, near Stroud.

Farmer Advice

A figure of new measures are being implemented, aimed at helping owners of bTB restricted herds to maintain their businesses and avoid some of the practical problems created by movement controls. Farmers wishing to find out more about bTB should contact their local AHO. Government-funded advice (based on the latest scientific evidence) is being developed, covering veterinary, biosecurity, and business issues. FCN agents will provide practical support, sign-post businesses to sources of other more specialist advice, and for those in greatest financial need a dedicated FCN Business Support Group will advise farmers on their options.

PROTECTING CATTLE FROM LEAD POISONING ON FARMS

Lead poisoning can cost you money and kill your cattle. The highest incidence usually occurs immediately after turn out when animals discover lead-containing materials abandoned on the pasture. Accumulation of lead beyond legal limits renders meat, offal and milk unsafe and illegal to enter the food chain. Lead poisoning can also result in stunted animal growth, animal deaths, increased birth defects and infertility, decreased productivity, loss of market value and disposal costs for dead animals and vet fees.

Causes of Lead Poisoning on Farms

The most frequent causes of lead poisoning in cattle are:

- Flaking high lead paint-mainly calves
- Vehicle and electric fence batteries-*e.g.* battery remains accidentally mixed with animal feed, or batteries fly-tipped on farm land
- High lead soils-usually arising from historic mining and smelting operations, land erosion-especially by water courses but occasionally landslips
- Ash from fires in which lead materials were burned such as painted woodwork, leaded building materials, putty, wiring
- Lead shot from shooting which can be eaten with soil uptake and can also contaminate certain crops, especially maize and end up in silage

Withdrawal Periods and Offal Removal

A 16-week withdrawal period before slaughter is usually sufficient but for lead which is retained in the stomach this can extend for several years. Some animals may show no signs of poisoning but have lead residues in milk, offal and meat. Offal tends to have higher levels of lead for longer periods than meat or milk. Animals and/or their produce may need to be tested to investigate whether lead residues are present and also to monitor whether a withdrawal period has been adequate or whether offal should be removed after slaughter.

How to Avoid Lead Contamination on Farm

Present are several steps you can take to protect your cattle and the human food chain from lead contamination. You should:

- Check your fields and barns for vehicle batteries, building materials, flaking lead paint, putty, lead flashing
- Remove or fence off fly-tipped material
- Prevent access to burnt out cars and old machinery that might contain lead
- Prevent cattle access to bonfire ash.

On farms with high lead soils, you should:

- Keep your cows' soil consumption as low as possible
- Avoid waterlogged land and poached land for grazing
- Avoid overgrazing and maintain adequate sward height
- Calibrate cutters when making silage to minimise soil uptake
- Flatten any molehills prior to cutting grass for silage
- Use mains water or tested borehole water rather than natural run-off water from high lead soils.

Suspect Lead Poisoning in Cattle

Some of your cows are contaminated with lead, you must:

- Remove your cattle from the affected area (pasture, pen or yard) immediately
- Consult your vet
- Confirm the cause of disease and if it is lead poisoning then investigate the source.

Testing for Lead in Cattle

Advice and testing for lead in cattle, produce or soil is available via your vet and from your regional Veterinary Laboratories Agency (VLA).

CATTLE FEED

Safe, good-quality foodstuffs are essential to maintain both human and animal health. Feed contaminants such as lead or antimicrobial residues-or biological agents such as botulism-may cause disease in cattle and make their produce unsuitable for human consumption.

Animal by-products as Foodstuffs

Some former foodstuffs-food previously intended for human consumption-as livestock feed, subject to the animal by-products regulations. Former foodstuffs that can be fed to cattle include:

- Milk and milk-based products
- Biscuits
- Bakery waste
- Pasta
- Chocolate
- Sweets and similar products

Such products can contain rennet, melted fat, milk or eggs, as long as they are not the main ingredients. You must ensure that the products are not contaminated by meat or other animal products before feeding them to cattle. You must not feed meat, fish and most other products of animal origin to ruminants, pigs or poultry, or allow them access to such material. Cattle should not be fed any processed or unprocessed catering waste, even if it comes from vegetarian restaurants and kitchens.

Transmissible Spongiform Encephalopathies (TSEs)

There are specific rules to prevent the spread of Bovine Spongiform Encephalopathy (BSE) and other TSEs through the use of animal by-products in animal feed.

Zoonoses

Zoonoses are defined diseases and infections that can be transmitted between vertebrate animals and man. Infection may occur through a bacterium, virus, fungus, parasite, or other communicable agent. Many of these can occur in cattle feed or milk and beef products from infected animals. Notifiable zoonoses include:

- Anthrax
- BSE
- Brucellosis
- Rift Valley Fever
- Bovine tuberculosis.

HORMONAL TREATMENTS AND ANTIBIOTICS FOR CATTLE

Due to concerns about the potential risk to humans, the use of hormonal growth promoters for livestock. Antibiotic growth-promoting feed additives have also been phased out-because of concerns about the potential spread of antibiotic resistance. If you keep farmed animals there are restrictions on the use of treatments that:

- Act as beta-agonists
- Have hormonal actions
- Reduce production of thyroid hormones.

Restricted treatments include:

- Phenylbutazone
- Tetracyclines

- Sulphonamides
- Nalidixic acid
- Fusidic acid
- Florphenicol
- Ceftriaxone.

Restricted antibiotic substances include:

- Chloramphenicol
- Ampicillin
- Amoxycillin/clavulanate
- Neomycin
- Gentamicin
- Enrofloxacin
- Tiamulin.

You must:

- Observe the relevant withdrawal period-*i.e.* the period between the end of treatment and the slaughter of the animal-if your food-producing animals have been given any of the restricted substances
- Keep veterinary medicinal records relating to restricted substances available to the competent authority on request.

You must not:

- Give food-producing animals restricted substances unless with any permitted exceptions
- Use substances containing oestradiol 17ß or its ester-like derivatives
- Use substances that contain beta-agonists to slow a labour-tocolysis-in cows when calving
- Use substances containing hormones or thyroid hormone-reducing actions unless prescribed by your veterinary surgeon
- Have food-producing animals on your farm that have been given any restricted substance, unless there are permitted exceptions
- Send animals to slaughter that have been given any restricted substances, unless there are permitted exceptions
- Sell meat or any other animal product that has been given a restricted substance.

8

Animal Cell Cultures

Animal cell cultures have been and are being used to generate valuable products based on their own genetic information or due to genes transferred into them (transgenes) using recombinant DNA technology. On the other hand, biotechnological approaches are used either to rapidly multiply animals of desired genotypes or to introduce specific alterations in their genotypes to achieve certain useful goals. To achieve the latter more efficiently, as well as to assist in conventional breeding efforts, the entire genomes of animals are being characterized using biotechnology tools. These activities have been arbitrarily grouped under animal biotechnology since they either utilize animal cells to generate products or apply biotechnological tools to enhance the usefulness of animals to human welfare.

WHAT IS ANIMAL CELL CULTURE?

Cell culture refers to the removal of cells from an animal or plant and their subsequent growth in a favourable artificial environment. The cells may be removed from the tissue directly and disaggregated by enzymatic or mechanical means before cultivation, or they may be derived from a cell line or cell strain that has already been established. Primary culture refers to the stage of the culture after the cells are isolated from the tissue and proliferated under the appropriate conditions until they occupy all of the available substrate (*i.e.*, reach confluence). At this stage, the cells have to be subcultured (*i.e.*, passaged) by transferring them to a new vessel with fresh growth medium to provide more room for continued growth.

After the first subculture, the primary culture becomes known as a cell line orsubclone. Cell lines derived from primary cultures have a limited life span (*i.e.*, they arefinite;), and as they are passaged, cells with the highest growth capacity predominate, resulting in a degree of genotypic and phenotypic uniformity in the population. If a subpopulation of a cell line is positively selected from the culture by cloning or some other method, this cell line becomes a cell strain. A cell strain often acquires additional genetic changes subsequent to the initiation of the parent line.

FINITE VS CONTINUOUS CELL LINE

Normal cells usually divide only a limited number of times before losing their ability to proliferate, which is a genetically determined event known assenescence; these cell lines are known as finite. However, some cell lines become immortal through a process called transformation, which can occur spontaneously or can be chemically or virally induced. When a finite cell line undergoes transformation and acquires the ability to divide indefinitely, it becomes a continuous cell line.

HISTORY OF ANCIMAL CELLS

Animal cells were first grown in vitro early in the twentieth century. In 1912, Alexis Carrel began growing bits of chick heart in drops of horse plasma. The cells at the edge of the explant divided and grew out of the plasma clot. The explants died within a few days, and Carrel reasoned that their death was due to exhaustion of nutrients. He found that cells from a given explant could be maintained indefinitely if they were periodically subdivided and fed with a sterile aqueous extract of whole chick embryos.

In the early 1950s, Earle (2) developed a technique for dissociating cells of a whole chick embryo from each other with trypsin. When this suspension of single cells was mixed with plasma and embryo extract and placed in a sterile glass container, the cells adhered to the glass and divided to form a primary culture. The primary culture contained a variety of cell types including macrophages, muscle fibers, etc. Some types of cells grew to a monolayer, a thin sheet of cells one layer in thickness, that covered the entire bottom surface of their culture vessel and then stopped dividing. The cells could then be redispersed with trypsin and planted in new culture vessels containing fresh medium. These secondary cultures contained fewer cell types than did the primary cell cultures, as many of the differentiated primary cells had not divided and were diluted out upon transfer. Often, secondary cultures were composed entirely of spindle-shaped cells called fibroblasts because of their similarity to cultured connective tissue. Cells derived from kidneys and from certain carcinomas had a polygonal appearance in culture. Because of their tissue of origin, they and other cells with similar morphology are called epithelial.

Purpose of Cell Culture

Cell culture is essential for growth of viruses because they are obligate intracellular parasites; they cannot replicate in any cell-free medium, and they require living cells from a suitable host within which to multiply. Animals such as mice and embryonated avian eggs may be used for the propagation of viruses, but for various reasons (time, cost, uniformity, ease of handling, animal welfare, etc.), the propagation of most viruses in cultured living cells is the method of choice today. Cell culture is also used in recombinant DNA technology and many in vitro assays. In fact most of what we know about the function and regulation of mammalian genes comes from studying genes transfected into cell culture (3).

Theory of Cell Culture

Cells may be grown in vitro in several ways. Organ cultures, if carefully handled, maintain their original architecture and functions for several days or sometimes weeks. Slices of organs (which are actually tissue cultures) consisting of respiratory epithelium have been used to study the histopathogenesis of infection by certain respiratory viruses that can only be grown outside their natural host in complex organized tissue. The term tissue culture was originally applied to explants of tissue embedded in plasma. The term subsequently became associated with the culture of cells in general and is now obsolete in its original sense. Cell culture is the term most correctly used today. It refers to tissue dissociated into a suspension of single cells by both mechanical and enzymatic means. After being washed and counted, the cells are diluted in growth medium and allowed to settle onto the flat bottom surface of a glass or specially treated plastic container. Most types of cells adhere quickly, and under optimum conditions they will undergo mitosis and cell division about once a day until the surface is covered with a confluent cell monolayer. Such cells constitute a primary culture. These cells are usually removed from the surface on which they are growing with a combination of trypsin and a chelating agent such as EDTA, counted, diluted, and replated in new containers with fresh medium. This is called a secondary culture.

There are three main types of cultured cells. The difference in these types lies in the number of times the cells can divide.

Primary cell cultures. When cells are taken freshly from animal tissue and placed in culture, the cultures consist of a wide variety of cell types, most of which are capable of very limited growth in vitro, usually fewer than ten divisions. These cells retain their diploid karyotype, i.e., they have the chromosome number and morphology of their tissues of origin. They also retain some of the differentiated characteristics that they possessed in vivo. Because of this, these cells support the replication of a wide range of viruses. Primary cultures derived from monkey kidneys, mouse fetuses, and chick embryos are commonly used for diagnostic purposes and laboratory experiments.

Diploid cell strains. Some primary cells can be passed through secondary and several subsequent subcultures while retaining their original morphological characteristics and karyotype. Subcultures will have fewer cell types than primary cultures. After 20 to 50 passages in vitro, these diploid cell strains usually undergo a crisis in which their growth rate slows and they eventually die out. Diploid strains of fibroblasts derived from human fetal tissue are widely used in diagnostic virology and vaccine production.

Continuous cell lines. Certain cultured cells, notably mouse fetal fibroblasts, kidney cells from various mammalian species, and human carcinoma cells, are able to survive the growth crisis and undergo indefinite propagation in vitro. After several passages, the growth rate of the culture slows down; then isolated colonies of cells begin to grow more rapidly than diploid cells, their karyotype becomes abnormal (aneuploid), their morphology changes, and other poorly understood changes take place that make the cells immortal. The cells are now "dedifferentiated," having lost the specialized morphology and biochemical abilities they possessed as differentiated cells in vivo. Continuous cell lines such as KB and HeLa, both derived from human carcinomas, support the growth of a number of viruses.

These lines and others derived from monkey kidneys (e.g., Vero), mouse fetuses (L929), and hamster kidneys (BHK) are widely used in diagnostic and experimental virology. Continuous cell lines have been established from many types of vertebrate and invertebrate animal tissues and are available from the American Type Culture Collection.

The development of antibiotics during World War II simplified long-term animal cell culture by minimizing the problems of bacterial and fungal contamination. Eagle made another important discovery in the 1950s when he determined the minimal nutritional requirements of cultured cells (1). He began by showing that HeLa and mouse L cells would grow in a mixture of salts, amino acids, vitamins and cofactors, carbohydrates, and horse serum. By eliminating one component at a time, he then determined which nutrients were essential for cell growth. His minimum essential medium (MEM) contains 13 amino acids (human tissue in vivo requires only eight), 8 vitamins and cofactors, glucose as an energy source, and a physiological salt solution that is isotonic to the cell. The pH is maintained at 7.2 to 7.4 by $NaHCO_3$ in equilibrium with CO_2. The pH indicator phenol red is usually incorporated into the medium; it turns red-purple if the medium is basic, yellow if the medium is acidic, and remains red-orange if the pH is in the right range.

Serum in concentrations of 1 to 10% must be added to the medium to provide the cells with additional poorly defined factors, without which most cells will not grow. Most mammalian cells are incubated at 37oC; avian, reptilian, and arthropod cells may grow best at higher or lower temperatures. If cells are grown in vessels open to the atmosphere, their incubator must be humidified and contain an increased CO_2 concentration. Some nonvolatile phosphate or substituted sulfonic acid buffers (HEPES, TES) eliminate the requirement for incubators to be gassed with CO_2. Tightly sealed containers also eliminate the need for CO_2. Other conditions required for growth of cells in culture are a negatively-charged solid surface (some tumor cells can be adapted to grow in suspension) and a minimum concentration of 2 x 105 to 5 x 105 cells per ml of medium or cm2 of growth area.

With the advent of cell culture, many animal viruses have been propagated in vitro, and hundreds of previously unknown viruses have been isolated and identified.

The discovery of the adenoviruses, echoviruses, and rhinoviruses, for example, is directly attributable to the use of cultured cells, as is the revolution in the diagnosis of viral diseases and the development of poliomyelitis, measles, rubella, and other vaccines.

CULTURE CONDITIONS

Culture conditions vary widely for each cell type, but the artificial environment in which the cells are cultured invariably consists of a suitable vessel containing the following:

- A substrate or medium that supplies the essential nutrients (amino acids, carbohydrates, vitamins, minerals)
- Growth factors
- Hormones
- Gases (O_2,)H_2O
- A regulated physico-chemical environment (pH, osmotic pressure, temperature).

Most cells are anchorage-dependent and must be cultured while attached to a solid or semi-solid substrate (adherent or monolayer culture), while others can be grown floating in the culture medium (suspension culture).

CRYOPRESERVATION

If a surplus of cells are available from subculturing, they should be treated with the appropriate protective agent (*e.g.*, DMSO or glycerol) and stored at temperatures below –130°C (cry preservation) until they are needed.

MORPHOLOGY OF CELLS IN CULTURE

- *Fibroblasti:* (or fibroblast-like) cells are bipolar or multipolar, have elongated shapes, and grow attached to a substrate.
- *Epithelial*: like cells are polygonal in shape with more regular dimensions, and grow attached to a substrate in

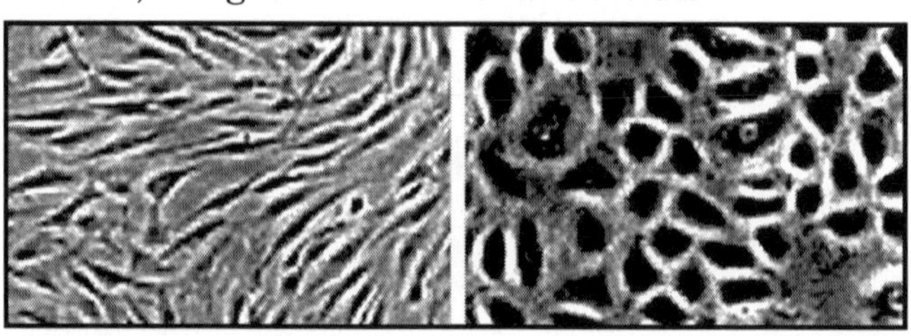

- *Lymphoblast:* like cells are spherical in shape and usually grown in suspension without attaching to a surface.

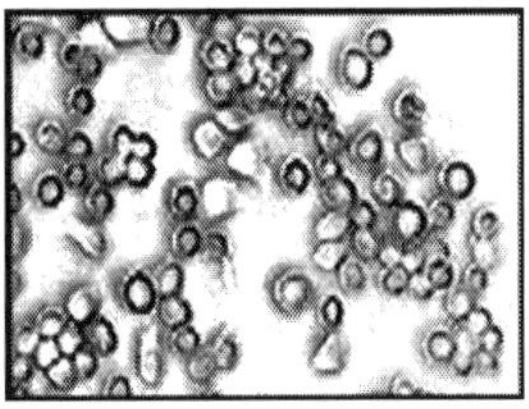

APPLICATIONS

Cell culture is one of the major tools used in cellular and molecular biology, providing excellent model systems for studying the normal physiology and biochemistry of cells (*e.g.*, metabolic studies, aging), the effects of drugs and toxic compounds on the cells, and mutagenesis and carcinogenesis. It is also used in drug screening and development, and large scale manufacturing of biological compounds (*e.g.*, vaccines, therapeutic proteins). The major advantage of using cell culture for any of these applications is the consistency and reproducibility of results that can be obtained from using a batch of clonal cells.

VARIETIES OF CELLS

Great varieties of cells are found in the various mature tissues of the higher animals, all of which have passed through the stage of being a simple nucleated mass of protoplasm in the earlier periods of their development. All cells may then be divided into two chief types, the indifferent and the differentiated. Under the category of indifferent cells may be placed all such as retain the characters of the first embryonic cells, and have not acquired any special structure or property by which they can be distinguished from the simplest form. Such cells are the only ones in the early stages of the embryo. In the adult tissues they also occur, having various duties to perform. They are found in the blood and lymph, and scattered throughout the tissues. They are without a cell wall, and have no special contents to mark their function.

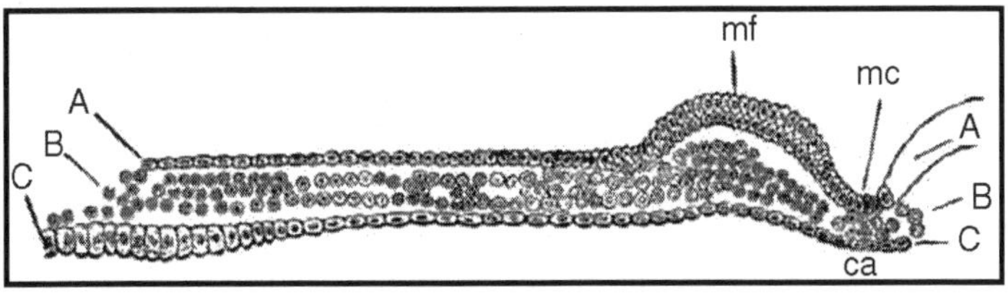

Fig. Transverse Section of Blastoderm, Showing the Elements in the Earlier Stage of the Development. A, Epiblast; B, Mesoblast; C, Hypoblast.

Among the differentiated cells we find many special characters, adapting them to certain special duties, for all these cells are modified from the original type and applied to the performance of some special function.

Space prevents even a short enumeration of the varieties of cells met with in the tissues of plants, where they not only carry on the active functions of the organism, but also form the supporting structures. The differentiation of a cell is accomplished by its protoplasm, which forms new structural parts, and itself sometimes seems to diminish in quantity until an element is produced in which there may be no protoplasm recognizable.

We find, then, matured and differentiated cells which vary:

- In shape, being spherical, flattened, fusiform, stellate, etc.
- In size.
- In their mode of connection.

Cells may also be classified according to their function, *e.g.*, Glandular, Nervous, etc., and the greater portion of the following pages will be devoted to the functions of these various forms of cells. So long as a cell remains in its indifferent stage it possesses the properties of ordinary protoplasm only; but by its further development it acquires special properties not common to all protoplasm.

These properties may or may not be accompanied by structural change. Thus the protoplasm of a gland cell differs in little from that of any other cell except in the capabilities of its nutritive changes and its chemical products; while on the other hand, those epithelial cells which form the outer layer of the skin lose their protoplasmic characters and are completely modified in structure.

Tissue Differentiation

The first stage in the existence of any organism, from the simplest form of plant to man, consists of a single cell (in animals called the ovum or egg), which differs in no essential points of structure from an ordinary cell.

There is moreover a class of organisms in which the individuals never go beyond this stage, but pass their entire lifetime in the state of a simple unicellular organism. The individuals composing this group (Protista), though insignificant in point of size, may vie with the higher plants and animals in number, species, and variety of form, so that they might well be placed in a kingdom by themselves (as has been proposed), apart from the vegetable and animal kingdoms.

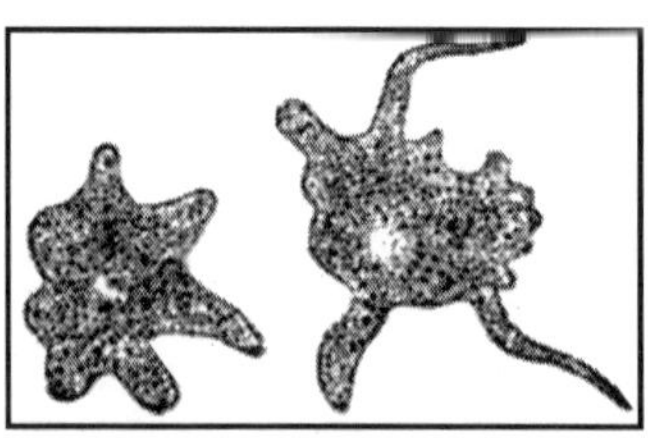

Fig. Unicellular Organism, Small Amoeba.

The group of these organisms which most resembles animals, is called Protozoa, and is divided from other animal forms by the manner of development of the ovum of the latter, which divides into cells that subsequently become differentiated into tissue. This group is called the Metazoa.

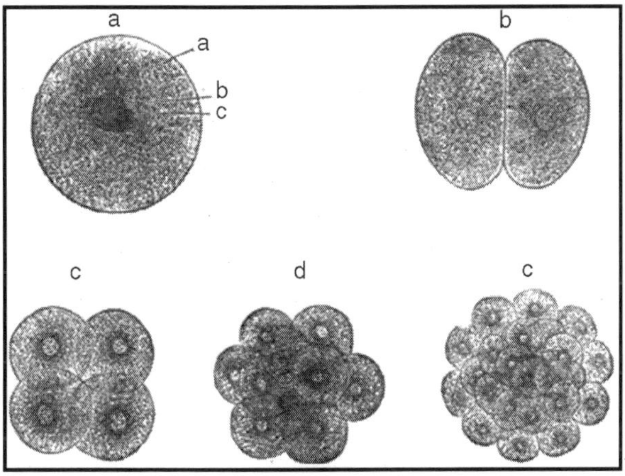

Fig. Stages in the Division of the Egg Cell (ovum), showing the Production of a Multiple Mass by Division. (Gegenbauer).

In the Protozoa the ovum never divides, the animal always remaining a single cell. On the contrary, the ovum of the Metazoa changes its characters during its development. At first possessing a stage common to both divisions, *viz.*, a single cell, it soon passes through rapid stages of cell proliferation, and is converted into a multiple mass, the mulberry stage or Morula. The cells forming this Morula stage approach the periphery of the mass, where they arrange themselves in two layers, and form a cavity in the centre. This is known as the Gastrula stage. Following, then, this cell multiplication, we find a qualitative differentiation of the cells, by which certain groups of cells assume special peculiarities, fitting them for some specific duty.

Thus we arrive at the production of special textures and organs such as are met with in the higher animals, and which are necessary for the efficient discharge of the various functions carried on during their lives. The division of the original mass of indifferent cells into two layers of special cells is the first step towards tissue differentiation, and in some animals is the only one arrived at in their life history, throughout which they remain a simple sac made up of an external layer, Ectoderm, and an internal layer, Endoderm.

The groups of cells forming the outer and inner layers of this stage of development, not only form the primitive tissues, but also represent the first appearance of organs or parts with a specific function. The external or ectodermic layer is the supporting, protecting, motor and respiratory organ, while the inner or endo-dermic layer is devoted to a primitive form of digestion, preparing the food for assimilation, and generally presiding over the nutrition of the body.

Although this sac-like (Gastrula) stage is supposed to have formed a step in the life history of nearly all animals, yet it forms a less striking part in the development of the individuals as we ascend the scale, and in the higher animals no such stage has been recognized. In the Vertebrates, the germ cells derived from the ovum are from an early period divided into three distinct layers, as those which correspond to the Ectoderm and Endoderm' of the lower organisms form between them a third layer or Mesoblast. From these germinal layers all the organs and tissues of the body are subsequently evolved. In embryological language the three primitive layers are called Epi-, Meso, and Hypo-blast. Thus it can be seen that, as we can compare the primitive unicellular state of the lowest animals with the first egg-cell stage of existence of the highest animals, so we can compare all the steps of tissue and organ differentiation as we trace them in the embryo of a mammal, with the steps of elaboration in organic and textural parts that we find in ascending the scale of animal life.

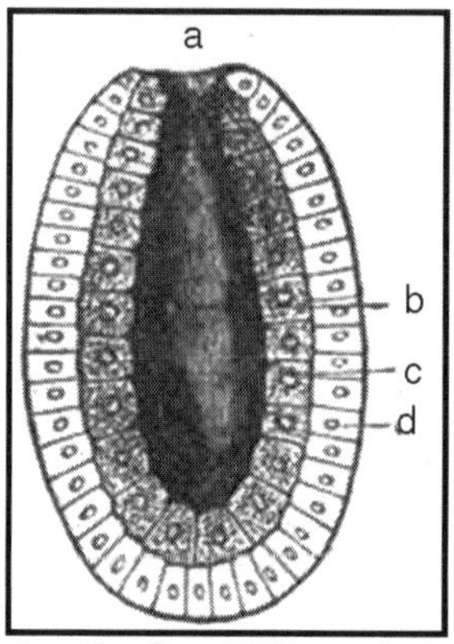

Fig. Diagram showing the first Differentiation of the Organism into an-External and Internal layer, (a) Mouth, (6) Alimentary Cavity, (d) Ectoderm, (c) Endoderm. (Gegenbauer).

The history, then, of the development of any mammal from a single cell or egg to the complex adult individual, is analogous with the more protracted history of the evolution of the animal kingdom from the Protista upward. It is impossible to separate the differentiation of tissues and organs, or to say which is of older date in the history of animal evolution.

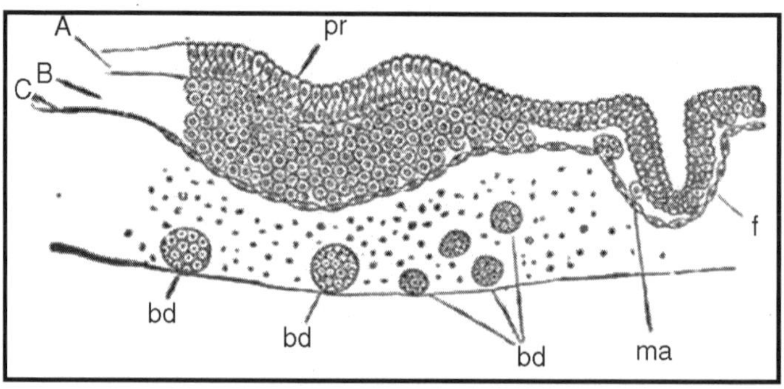

Fig. Transverse Section of Blastoderm of Chick. A. Epiblast. B. Mesoblast. C. Hypoblast. pr. Primitive Groove.

Even in unicellular animals, where we have no trace of tissue difference, there being only one cell, we have a distinct foreshadowing of organ and functional differentiation. And in creatures made of many parts, the same cells have several duties to perform. But when an aggregation of specialized cell units exists, it may be said to be a tissue. If these cells have no very special characteristic, then the tissue may be called primitive or embryonic. But, as has just been stated, the aggregation of embryonic cells - in the higher forms of life—have special characters from the very first, which mark them off from one another as destined for different functions.

TECHNIQUES AND PROCEDURE FOR ORGAN CULTURE

In order to optimize the nutrient and gas exchanges, the tissues are kept at gas limited interface using the support material which ranges from semisolid gel of agar, clotted plasma, micropore filter, lens paper, or strips of Perspex or plexiglass. The organ cultures can also be grown on top of a stainless steel grid. Another popular choice for growing organ cultures is the filter-well inserts. Filter-well inserts with different materials like ceramic, collagen, nitrocellulose are now commercially available. Filter well inserts have been successfully used to develop functionally integrated thyroid epithelium, stratified epidermis, intestinal epithelium, and renal epithelium.

The procedure for organ cultures has the following steps:

- The organ tissue is collected after the dissection.
- The size of the tissue is reduced to less than 1mm in thickness.
- The tissue is placed on a gas medium interface support.
- Incubation in a CO_2 incubator.
- M199 or CMRL 1066 medium is used and changed frequently.
- The techniques of histology, autoradiography, and immunochemistry are used to study the organ cultures.

THE ADVANTAGES OF ORGAN CULTURE

The organ cultures can be used to study the behaviour of an integrated tissue in the laboratory. It provides an opportunity to understand the biochemical and molecular functions of an organ/tissue.

LIMITATIONS OF ORGAN CULTURE

It is a difficult and expensive technique. The variations are high with low reproducibility. For each experiment, a new or fresh organ is needed as organ cultures are not propagated.

HISTOTYPIC CULTURES

Using histotypic culture, it is possible to use dispersed monolayers to regenerate tissue like structures. It the growth and propagation of cell lines in three-dimensional matrix to high cell density that contributes to this. The techniques used in histotypic cultures are:

Gel and Sponge Technique

In this method, the gel (collagen) or sponges (gelatin) are used which provides the matrix for the morphogenesis and cell growth. The cells penetrate these gels and sponges while growing.

Hollow Tibers Technique

In this method, hollow fibres are used which helps in more efficient nutrient and gas exchange. In recent years, perfusion chambers with a bed of plastic capillary fibres have been developed to be used for histotypic type of cultures. The cells get attached to capillary fibres and increase in cell density to form tissue like structures.

Spheroids

The re-association of dissociated cultured cells leads to the formation of cluster of cells called spheroids. It is similar to the reassembling of embryonic cells into specialized structures.

The principle followed in spheroid cultures is that the cells in heterotypic or homotypic aggregates have the ability to sort themselves out and form groups which form tissue like architecture. However, there is a limitation of diffusion of nutrients and gases in these cultures.

Multicellular Tumour Spheroids

These are used as an in vitro proliferating models for studies on tumour cells. The multicellular tumour spheroids have a three dimensional structure which helps in performing experimental studies related to drug therapy, penetration of drugs besides using them for studying regulation of cell proliferation, immune response, cell death, and invasion and gene therapy. A size bigger than 500 mm leads to the development of necrosis at the centre of the MCTS.

The monolayer of cells or aggregated tumour is treated with trypsin to obtain a single cell suspension. The cell suspension is inoculated into the medium in magnetic stirrer flasks or roller tubes.

After 3-5 days, aggregates of cells representing spheroids are formed. Spheroid growth is quantified by measuring their diameters regularly. The spheroids are used for many purposes. They are used as models for a vascular tumour growth. They are used to study gene expression in a three-dimensional configuration of cells. They are also used to study the effect of cytotoxic drugs, antibodies, radionucleotides, and the spread of certain diseases like rheumatoid arthritis.

ORGANOTYPIC CULTURES

These cultures are used to develop certain tissues or tissue models for example skin equivalents have been created by culturing dermis, epidermis and intervening layer of collagen simultaneously. Similarly models have been developed for prostrate, breast etc.

Organotypic culture involves the combination of cells in a specific ratio to create a component of an organ.

APPLICATIONS OF ANIMAL CELL CULTURE

Animal cells can be cultured after being transduced with nucleic acid(s). The cells in culture would behave in almost the same way as they would behave in an animal body. So this would help us to understand the changes in gene-expression due to the altered genotype. The animal cell cultures are used for a diverse range of research and development.

These areas are:

- Production of antiviral vaccines, which requires the standardization of cell lines for the multiplication and assay of viruses.
- Cancer research, which requires the study of uncontrolled cell division in cultures.
- Cell fusion techniques.
- Genetic manipulation, which is easy to carry out in cells or organ cultures.
- Production of monoclonal antibodies requires cell lines in culture.
- Production of pharmaceutical drugs using cell lines.
- Chromosome analysis of cells derived from womb.
- Study of the effects of toxins and pollutants using cell lines.
- Use of artificial skin.
- Study the function of the nerve cells.

SOMATIC CELL FUSION

One of the applications of animal cell culture is the production of hybrid cells by the fusion of different cell types.

These hybrid cells are used for a the following purposes:

- Study of the control of gene expression and differentiation,
- Study of the problem of ' malignancy',
- Viral application,
- Gene mapping,
- Production of hybridomas for antibody production.

In 1960s, in France for the first time, the hybrid cells were successfully produced from mixed cultures of two different cell lines of mouse. Cells growing in culture are induced by some of the viruses such as 'Sendai virus' to fuse and form hybrids. This virus induces two different cells first to form heterokaryons. During mitosis, chromosomes of heterokaryon move towards the two poles and later on fuse to form hybrids. It is important to remove the surface carbohydrates to bring about cell fusion. Some other chemicals like polyethylene glycol also induce somatic cell fusion. Many commercial proteins have been produced by animal cell culture and there medical application is being evaluated.

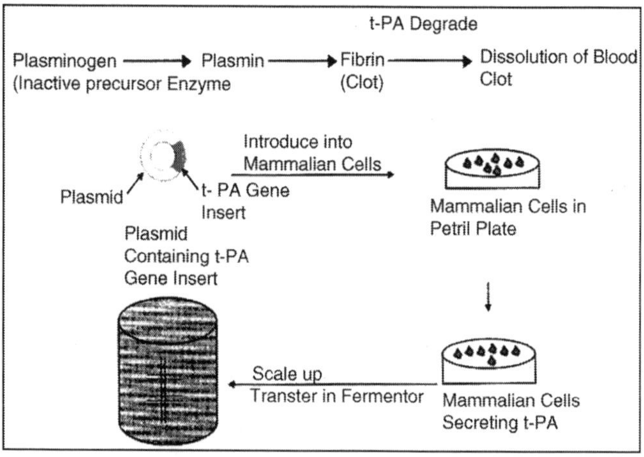

Fig: Showing the Production T-PA

Tissue Plasminogen activator (t-PA) was the first drug that was produced by the mammalian cell culture by using rDNA technology. The recombinant t-PA is safe and effective for dissolving blood clots in patients with heart diseases and thrombotic disorders.

BLOOD FACTOR VIII

Haemophilia A is a blood disorder which is a sex-linked genetic disease in humans. The patients suffering from Haemophilia A lack factor VIII, which plays an important role in the clotting of blood.

This factor VIII is secreted by a gene present on X-chromosome but this gene undergoes mutations in people suffering from Haemophilia. Current therapy for this disease is the transfusion of blood factor VIII into patients. Using rDNA technology, Factor VIII has been produced from mammalian cell culture *e.g.* Hamster kidney cell.

ERYTHROPOIETIN (EPO)

The EPO is a glycoprotein consisting of 165 amino acids and is formed in the foetal liver and kidneys of the adults. It causes proliferation and differentiation of progenitor cells into the erythrocytes (erythroblasts) in the bone marrow. Erythropoietin is hormone-like in nature and is released by the kidney under hypoxic or anoxic conditions caused by anaemia.

Amgen Inc. holds US patent for preparation of, eErythropoietin, by recombinant method using Chinese Hamster Ovary cell lines. Erythropoietin (EPO) is a hormone-like substance released by the kidney under hypoxic or anoxic conditions caused by anaemia. r-HUEPO- recombinant human erythro-protein has been effectively used to treat anemia associated with AIDS, renal failure etc.

THE PRODUCTION OF MONOCLONAL ANTIBODIES USING HYBRIDOMA TECHNOLOGY

Antibodies are proteins synthesized in blood against antigens and are collected from the blood serum. The antibodies, which are heterogenous and non specific in action are called polyclonal antibodies. If a specific lymphocyte, after isolation and culture in vitro becomes capable of producing a single type of antibody bearing specificity against specific antigen, it is known as monoclonal antibody.

The monoclonal antibodies are used in the diagnosis of diseases because of the presence of desired immunity. However, these antibody secreting cells cannot be maintained in culture. It was observed that the myeloma cells (bone marrow tumour cells due to cancer) grow indefinitely and also produce immunoglobulins which are infact monoclonal antibodies. In 1974, George Kohler and Milstein isolated clones of cells from the fusion of two parental cell lines–lymphocytes from spleen of mice immunized with red blood cells from sheep and myeloma cells. These cells were maintained in vitro and produced antibodies.

The hybrid cells maintained the character of lymphocytes to secrete the antibodies, and of myeloma cells to multiply in culture. These hybrid cell lines are called "Hybridoma" and are capable of producing unlimited supply of antibodies.

Hybridoma are obtained by using an antibody producing lymphocytes cell and a single myeloma cell. Monoclonal antibodies bind very specifically to an epitope (specific domains) on an antigen and by using them it is possible to detect the presence of specific antigens.

The Monoclonal antibodies are used for the treatment of patients with malignant leukaemia cells, B cell lymphomas and allograft rejection after transplantation. CD3 is an antigen present on the surface of mature T- cells lymphocytes.

If T- cell population is depleted or controlled, the transplanted organ will not be rejected. An antibody that acts against CD3 surface antigen of T-cells is called OKT3 *i.e.* anti-CD3 Moab. OKT3 is a monoclonal antibody which has been licensed for clinical use for the treatment of acute renal allograft rejection. OKT3 removes antigen bearing cells from circulation thereby helps in accepting the graft.

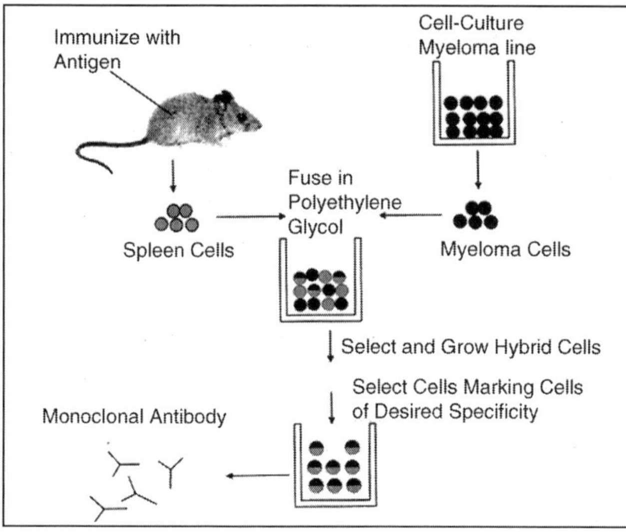

Fig: Showing The Steps Involved in the Production of Monoclonal Antibodies

When Monoclonal antibodies are used as enzymes using the technique of enzyme engineering, then they are called abzymes. Using animal cell cultures, it is also possible to produce Polyclonal Antibodies. Polyclonal antisera are derived from many cells therefore contains heterogeneous antibodies that are specific for several epitopes or an antigen.

VIRUS VACCINES IN ANIMAL CELL CULTURE

Animal cell cultures are used to produce virus vaccines, as well as a variety of useful biochemical's which are mainly high molecular weight proteins like enzymes, hormones, cellular biochemical's like interferon's, and immunobiological compounds including monoclonal antibodies. Animal cells are also good hosts for the expression of recombinant DNA molecules and a number of commercial products have been/are being developed. Initially, virus vaccines were the dominant commercial products from cell cultures, but at present monoclonal antibody production is the chief commercial activity.

It is expected that recombinant proteins would become the prime product from cell cultures in the near future. Transplantable tissues and organs are another very valuable product from cell cultures. Artificial skins are already in use for grafting in burn and other patients, and efforts are focused on developing transplantable cartilage and other tissues.

A greater detail of cell culture products is provided under the following heads:

- Vaccines,
- Interferon,
- Monoclonal antibodies,
- Hybrid antibodies
- Recombinant proteins.

A vaccine is a preparation containing a pathogen (disease producing organism) either in attenuated or inactivated state. This preparation is introduced into an individual to induce adequate antibody production against the pathogen in question so that the individual becomes protected against infection, at a later date, by that pathogen. The introduction of a vaccine in an individual is called vaccination or immunization as it leads to the development of immunity in the vaccinated individuals to the concerned pathogen. The immunity is induced by the antigens of pathogen origin and present in the vaccine.

Any molecule that induces production of antibodies specific to itself when introduced in the body of an animal is called antigen. Usually the antigenic function is confined to a rather small portion of the antigen molecules; such a region is known as antigenic determinant or epitope. A B-lymphocyte cell becomes predetermined, during differentiation, to produce antibodies of a single specificity. Ordinarily in an animal, the frequency of B-lymphocytes capable of producing antibodies of a given specificity, *i.e.*, against a single epitope, is very low. (However, the B-lymphocyte population of an animal produces antibodies against an amazingly large number of epitopes or antigenic determinants).

When an antigen enters into the body of an animal, its antigenic determinant ultimately binds to the specific receptors present on the surface of those Blymphocytes which produce antibodies specific to this antigen. This binding stimulates a rapid proliferation of these Blymphocytes so that the proportion of such cells increases drastically; this is called clonal selection (proposed by N. Jerne). As a consequence, the concentration in the blood serum of antibodies (antibody titre) specific to the concerned antigen also increases dramatically. This phenomenon is the basis of immunization.

When an individual is vaccinated, the antigens of pathogen origin stimulate antibody production against themselves. This increase in antibody production takes some time, but since the vaccine does not have virulent pathogens there is no danger of disease development. (If the pathogen present in the vaccine were virulent/alive, vaccination would produce disease before the antibody production could be sufficiently increased). When live virulent form of the same pathogen later enters into the system of an immunised animal, the high level of antibodies specific to the pathogen (present in the animal as a result of immunization) inactivate the pathogen, and thereby protect the animal against the disease.

The various vaccines can be grouped into two categories:

- Vaccines containing killed or inactivated pathogenes, *i.e.*, most bacteria vaccines and some virus vaccines (*e.g.*, influenza virus inactivated by formalin, rabies virus inactivated by phenol or ppropiolactone),
- Those containing live but attenuated pathogens, *e.g.*, most virus vaccines.

Attenuation means a drastic reduction in the virulence of a pathogen; this is achieved as follows.

- Several consecutive passages through an animal which is not the usual host of the pathogen, *e.g.*, small pox virus in calf.
- Several passages through cultured cells of the host, *e.g.*, rabies virus in human diploid cell culture, or of a different species, *e.g.*, rabies virus, yellow fever virus in chick embryo cell culture.
- Selection of less virulent strains of pathogens, *e.g.*, a mutant strain of polio virus.
- Treatment of the pathogen with some chemicals, *e.g.*, B.C.G. (Bacillus of Calmette Guerien) vaccine produced by culturing the bacteria on a medium containing bile.
- Culturing pathogens under unfavourable conditions like high temperature, *e.g.*, anthrax vaccine obtained by cultivation of the bacterium (Bacillus anthracis) at 40°-50°C.

In general, inactivation of viruses is always coupled with attenuation to minimize the accidental presence in vaccines of active virulent particles which could cause disease in the vaccinated individuals. The different vaccines differ in their composition, efficacy and the duration of effective protection to the vaccinated individuals.

Vaccines are one of the earliest examples of biotechnological intervention in human and animal health care. (The first case of vaccination was against rabies by Louis Pasteur on July 6, 1885 in case of a young boy bitten by a mad dog. Pasteur used the fluid from spinal cord of (another mad dog for vaccination.) Vaccines offer the cheapest and most effective protection against diseases, and for some diseases, *e.g.*, hepatitis B, AIDS etc., they are the only means of protection. The effectiveness of vaccines may be highlighted by the success of WHO (World Health Organisation) sponsored mass vaccination against small pox in completely wiping out this once ravaging disease from the face of earth.

The procedure of virus vaccine production using cell cultures is essentially and in simple terms as follows. The cells to be used as host are inoculated in culture vessels and, after suitable growth has occurred; the cultures are infected with the virus concerned and incubated. After an appropriate period of virus multiplication and release, the culture medium containing virus particles is collected and suitably processed.

In case of killedvirus vaccines, the vaccine is usually subjected to a concentration step after virus inactivation. Suitable stabilizers are added to prevent loss of potency and the vaccines are generally stored at low temperature till used. A rigorous quality control is maintained throughout the entire operation. A number of vaccines are produced in cultured animal cells; some of these are being produced in India but the procedure has several inherent problems.

- Rigorous quality control is essential particularly to exclude the possibility of presence of infectious agents in the materials and cells used. In addition,
- the vaccine may contain a rare virulent and active virus particle which may cause disease in the vaccinated individuals.

Therefore, alternative approaches of vaccine production and delivery have been/are being developed, *viz.*,

- Production of antigenic proteins or only their antigenic determinants in GEMS
- Use of DNA as vaccines
- Production of antigenic proteins in vegetables/fruits to minimise storage costs and problems.

Interferon's

Interferon's are proteins produced by a cell infected by a virus, and provide protection to other healthy cells from viruses. Interferon was discovered in 1957 when Issacs and Lindenmann observed that virus-free fluid obtained from cultured cells infected with virus protected other cells from virus infection. They called the substance present in these fluids, which interfered with virus infection, interferon.

There are three major types of interferons:

1. Interferon-a (INF-a; produced by leucocytes or white blood cells),
2. Interferon-|3 (INF-p; produced by fibroblasts),
3. Interferon-y (inf-y; produced by stimulated T-lymphocyte cells, hence also called immune interferon).

The mechanism of protection by interferons appears to be as follows. When interferon reacts with the interferon receptors of a cell, the cell enters in a state called interferon-induced antiviral state. In this state, a rapid degradation of mRNA occurs if the cell is infected by any virus. Interferon induces in cells the production of 2, 5-adenosine polymerase. When such a cell is infected by a virus, the 2, 5-A polymerase is activated to produce 2, 5 adenosine (2, 5-A), which in turn activates pre-existing but inactive molecules of ribonuclease-L. Activated ribonuclease-L degrades all mRNA (of host as well as virus origin) present in the cell bringing the protein synthesis to a halt in such cells.

This interferes with multiplication of the virus so that virus infection is either stopped or sufficiently slowed down to allow the production of adequate antibodies against the invading virus. The protection due to interferons is nonspecific in that interferon induced by any one virus will provide protection against all viruses. It is possible that interferons modify ribosomes so that they no longer translate viral mRNAs, although they are fully capable of translating mRNAs of the host origin. Interferons enhance the cytotoxic activity of natural killer cells (NK cells), which are a type of lymphocytes identifiable as large granular lymphocytes (LGL). NK cells are cytotoxic to some types of tumour cells. Interferons are known to inhibit growth of some types of tumours; most of these tumors are also responsive to chemotherapy. Interferons, therefore, have been employed in treatment of the responsive tumors despite their prohibitive cost (initially, $50 million/g of commercial product). Interferons are produced from human leucocytes isolated from donor blood and cultured in vitro, and from mouse fibroblast cultures.

The production scheme, in simple terms, is as follows. Large scale (1,000 1 to 10,000 1) cell cultures are infected with Sendai virus, and incubated for 24 h after which the supernatant (clear fluid) is collected, centrifuged, and used for interferon isolation. The amount of interferon recovered is relatively small (1 g interferon of low purity from leucocytes separated from blood of about 90,000 donors), and the normal leucocytes are difficult to culture preventing scaling up from relatively small inocula. These contributed to the enormous price of the product. In view of the value of and demand for interferons, intensive efforts were made to produce it in genetically engineered organisms, *e.g*., E. coli, yeast mammalian cell cultures and even in plants These efforts have drastically reduced the cost (to less than 10% of the initial price) and improved the purity (by several orders of magnitude) of the product.

Recombinant Proteins

Proteins produced by genes transferred into selected host cells by genetic engineering are called recombinant proteins since they are based on recombinant DNA technology. Recombinant proteins form an important component of biopharmaceuticals, *i.e*., biotechnology products having pharmaceutical applications. A large number (nearly 2 dozen) of recombinant proteins are being produced in mammalian cell cultures some of which, *viz*., human growth hormone (HGH), tissue plasminogen activator (tPA), erythropoietin and blood clotting factor VIII, are already in therapeutic use. The host cells used for large scale production of the various recombinant proteins are: Chinese hamster ovary line (CHO), baby hamster kidney line BHK, mouse mammary tumour line C127, mouse myeloma cell lines and mammalian cell lines.

ANIMAL CELL TECHNOLOGY: BECOMING MORE AND MORE IMPORTANT

Stem cells retain the capacity to self renew as well as to produce progeny with a restricted mitotic potential and restricted range of distinct types of differentiated cell they give rise to. The formation of blood cells also called haematopoiesis is the classical example of concept of stem cells. Indirect assay methods were developed to identify the haematopoietic stem cells. The process ofhaematopoeis is occurs in the spleen and bone marrow in mouse. In human beings about 100,000 haematopoietic stem cells produce one billion RBC, one billion platelets, one million T-cells, one million B cells per kg body weight per day. Several methods have been developed to study haematopoiesis and stem cells.

By definition animal cell technology is that part of biotechnology which is based on the use of animal (including human) cells, propagated in vitro, for the manufacture of biological molecules, e.g. biopharmaceuticals, and as toolbox in the discovery and testing of medicines.

The main field of application of animal cell technology is the medical and pharmaceutical arena. The members of ACTIP, the Animal Cell Technology Industrial Platform, want to support and reinforce further development of animal cell technology which will lead to increased utilization of animal and human cells to be propagated in vitro.

As early as the 19th century, animal cell technology already formed the basis for providing vaccines against small pox and rabies, although these vaccines were actually produced in living animals.

In the 20th century, knowledge in many areas of the life sciences increased tremendously. In the early 1950s this led to the Salk and Sabin processes of in vitro cultivation of cells for the production of highly effective vaccines against polio virus. Other vaccines against a wide variety of human and animal viral diseases like rubella, measles, foot and mouth disease etc. followed soon after. In addition, novel animal cell technologies made it possible to develop high throughput assays for the screening of new candidate pharmaceuticals and as diagnostic tools.

Breakthroughs in the 1970s and 1980s

A major breakthrough in the field of animal cell technology came in the late 1970s/early 1980s with the development of hybridoma and recombinant DNA technology.

The hybridoma technology makes it possible to elicit, select and produce, in vitro, so called monoclonal antibodies of consistent quality, high specificity and unlimited quantities, whereas previously antibodies could only be produced in life animals with a varying degree of quality.

Besides the more "classical" use of antibodies in passive immunization of patients against infectious diseases, especially monoclonal antibodies (Mab) are nowadays widely used as a therapeutic mean to treat cancer, chronic and autoimmune diseases as well as for diagnostic purposes like blood group typing and disease testing.

The development of recombinant DNA technology or heterogenous gene expression allows for e.g. the production of human proteins in animal cells at a relatively efficient production scale. Previously, many of these human proteins could not be produced or isolated at all.

Need for animal cells

The recombinant DNA technology - also called genetic engineering - makes it possible to transfer a gene from one species into cells of another, and to let these transformed cells produce the encoded gene product. For example, the encoding gene for human insulin has been successfully transfected into the easy to grow bacterium Escherichia coli. However, with more complex and larger gene products than insulin, it is no longer possible to use bacteria like E. coli as host cells for the production.

The reason is that these bacteria can not perform the complicated posttranslational processing steps such as glycosylation, folding and trimming, which is often a prerequisite to exert biological functions. Alternatively, such proteins can be expressed successfully and correctly in cultured animal cells. For example, the bioactivity of erythropoietin, tissue plasminogen activator or factor VIII is strongly dependent on a correct posttranslational processing. Development of novel cell lines has resulted in an increased frequency in which animal cells are being used.

In the early eighties the majority of biopharmaceuticals was produced in E. coli (86%), but nowadays the majority of biopharmeuticals are produced in animal cells, where Chinese Hamster Ovary (CHO) has become the gold standard. This development has mainly be driven by the production of therapeutic monoclonal antibodies.

Using animal cells as production factories, however, had the disadvantage that the production costs are relatively high in comparison with production in cultivated bacteria or yeast cells, and related to this, the relatively small production scale. Addressing this situation was and still is in the focus of development activities on animal cells. Aims herein are development of robust high yielding cell lines, improvement of process conditions and media, etc.

THERAPEUTICS: DEMAND FOR LARGE SCALE MANUFACTURING FACILITIES

At present, the majority of therapeutic biopharmaceuticals has been produced using animal cell technology and include proteins used for the treatment of cardiovascular diseases (tissue plasminogen activator: tPA, reteplase), cystic fibrosis (DNases), anemia (erythropoietin: EPO), haemophilia (coagulation factors VIII and IX), cancer and viral infections (interferons and interleukins), multiple sclerosis (interferon-beta2) and dwarfism (human growth hormone: hGH). Monoclonal antibodies are used for the treatment of a rapidly expanding list of diseases, e.g. to reduce the rejection of transplanted organs by the recipient (e.g. Simulect and Zenapax), to treat metastasized breast cancer (Herceptin), to treat non-Hodgkin lymphoma's (Rituxan), to inhibit allergic asthma (Xolair), and many others (see monograph on monoclonal antibodies). These different antibody indications have in common that effective patient dosing requires relatively large amounts of antibody-protein, which translates into a huge demand for large scale manufacturing capacity.

Cell and Gene therapy

Culturing of animal cells forms also an important basis for the new area of cell and gene therapy. This kind of treatment stands further for a change in therapy, away from large size (patient numbers) treatments to more specific even single patient interventions. Obviously this will have an influence on manufacturing conditions.

Cell therapy is the use of human cells to treat patients, such as autologous applications where cells from a patient are removed, enriched (e.g. stem cells) and/or modified and finally re-injected in the patient. Applications are tissue regeneration after injury, rebuilding the immune system after chemotherapy, and others. The field itself is currently developing and manufacturing and regulatory questions need to be addressed to shape and design this new area.

In contrast to direct treatment of patients with recombinant biopharmaceuticals, gene therapy is based on the delivery of the corresponding gene to a patient, in such a way that the patient may produce the functional gene product (i.e. the protein) intracellularly. In principle, such an approach is considered to result in long-lasting treatment or cure, e.g. modification of the patients genome.

For the delivery of relevant genes into specific tissues in patients recombinant viruses are being used. These viruses are genetically modified and produced in vitro in animal cells. Gene therapy is very promising since it may yield a number of advantages. When we consider for example patients suffering from a bleeding disorder such as Haemophilia A, these patients have a deficient gene encoding for clotting Factor VIII. Current treatment is based on the life long treatment with the recombinant protein Factor VIII. Gene therapy might make it possible to administer the gene encoding Factor VIII into the liver cells of these patients. These cells are the natural producer cells of Factor VIII and, thus, may behave naturally after transfection with the functional gene encoding Factor VIII.

Drug discovery

The evolution of the drug discovery and development process has been revolutionized by the implementation of high-throughput technologies, combinatorial chemistry and genomics-based target identification and validation. Animal and human cells have become an essential part of the cell-based assays and models, which form the basis of target identification and validation experiments, HT-screening of gene (sequence) libraries for lead finding and optimization purposes and HT-testing for ADME/tox of candidate drugs. Given the fact that these areas are still in rapid development, it will be obvious that improvement of cell culture procedures remains a prerequisite. In this respect, multidisciplinary R&D programs remain crucial to feed further progress in the tool box needed for efficient drug discovery.

Animal cells are cultivated in vitro by mimicking the same conditions as they have in their natural environment, the body. They need complex and balanced culture media which contain substrates such as glucose, amino acids, vitamins, salts, trace elements and others. Small scale cultures are performed in tissue culture flasks or roller bottles in a 37°C incubator.

Large scale cultivation for industrial purposes is mainly performed in bioreactors which are large stainless steel tanks with agitation to mix the cell suspension and aeration to supply oxygen. Alternatively, cells can also be grown in disposable bags, and for a large capacity one simply needs to add a larger number of bags. In these systems, cells grow, multiply and, in the case of recombinant cells, release human proteins into the culture medium from which the protein product can be purified, formulated and packaged. Currently, a major development is to design large scale cultivation procedures utilizing media free of serum additives or protein growth factors. Resulting protein free media or procedures are considered to be far more safe as to the transfer of biological contamination. Particularly, potential contamination of biopharmaceutical end products with agents that may transfer Bovine Spongiform Encephalitis (BSE) has been the major driver to develop such protein/serum free media and related cell culturing processes.

ANIMAL CELLS AND THEIR APPLICATIONS

The mammalian cells are genetically modified by introducing the genes needed for specific purposes such as production of specific proteins or to improve the characteristics of a cell line. The methods used to introduce the foreign genes/DNA into mammalian cells are: Electroporation, Lipofection, Microinjection and/or fusion of mammalian cells with bacteria or viruses.

After the integration of the foreign DNA into the mammalian cells, the transfected/transformed cells are selected by using suitable markers. Some of such markers in use are: Viral thymidine kinase, Bacterial dihydrofolate reductase, Bacterial neomycin phosphotransferase. It has been possible to overproduce several proteins in mammalian cells through genetic manipulations *e.g.* tissue plasminogen activator, erythropoietin, interleukin-2, interferon- beta, clotting factors VIII and IX, tumor necrosis factors. The recombinant mammalian cells are also conveniently used for the production of monoclonal antibodies.

MANIPULATION OF GENE EXPRESSION IN EUKARYOTES

The eukaryotic organisms have the capability to bring about the post-translational modifications such as glycosylation, phospho-rylation, proteolytic cleavage etc which ultimately helps in the production of stable and biologically active proteins. Due to these reasons the use of eukaryotic expression system is preferred however it is difficult to conduct experiments with eukaryotic cells. The introduction of a foreign DNA into animal cells is called transfection. The insert DNA in the eukaryotic cells may be associated with vector or integrated into the host chromosomal DNA.

Among the various hosts used for the expression of cloned genes, the common yeast Saccharomyces cerevisiae is the most extensively used. Besides this, the cultured insect cells are in use for expressing cloned DNAs. Baculoviruses exclusively infect insect cells. The DNA of these viruses encode for several products and their productivity in cells is very high to the extent of more than 10,000 times compared to mammalian cells. The baculoviruses not only carry a large number of foreign genes but can also express and process the products formed. By using baculovirus as an expression vector system, a good number of mammalian and viral proteins have been synthesized. The most commonly used baculovirus is Autographa californica multiple nuclear polyhedrosis virus (AcMNPV). It grows on the insect cell lines and produce high levels of polyhedrin or a recombinant protein.

The mammalian cell expression vectors are used for the production of specific recombinant proteins and to study the function and regulation of mammalian genes. However, large-scale production of recombinant proteins with engineered mammalian cells is costly. The mammalian vector contains a eukaryotic origin of replication from an animal virus such as Simian virus 40 (SV 40) and a prokaryotic origin of replication. It has a multiple cloning site and a selectable marker gene, both of which remain under the control of eukaryotic promoter and polyadenylation sequences. These sequences are obtained from either animal viruses (SV40, herpes simplex virus) or mammalian genes (growth hormone, metallothionein). The promoter sequences facilitate the transcription of cloned genes (at the multiple cloning site) and the selectable marker genes. On the other hand, the polyadenylation sequences terminate the transcription.

COLLECTION AND PURIFICATION PROCESS OF RECOMBINANT PROTEINS

As the recombinant proteins start accumulating in the host cells, it becomes important to collect and purify them. This is a tricky process since many times the recombinant protein is a foreign body for the host cells and the enzyme machinery of the host cell becomes activated to degrade the outside protein. One of the strategies adopted is the use of bacterial strains deficient in proteases or alternatively, the recombinant proteins are fused with the native host proteins. The fusion proteins are resistant to protease activity.

Sometimes, the foreign proteins accumulate as aggregates in the host organism which minimizes the protease degradation. The best way out is to quickly export and secrete out the recombinant proteins in to the surrounding medium. The recovery and the purification of foreign proteins is easier from the exported proteins. The efforts have been made to develop methods to increase the export of recombinant proteins.

Some of the species of the bacterium, Bacillus subtilis normally secrete large quantities of extracellular proteins. A short DNA sequence called signal sequence from such species is introduced into other B. subtilis. These bacteria produce recombinant DNA tagged with signal peptide, which promotes export and secretion. This signal peptide is removed after the purification of foreign protein. The techniques used for the purification of recombinant proteins from the mixture of secreted proteins are affinity tagging, immunoaffinity purification etc.

ORGAN CULTURE AND HISTOTYPIC CULTURES

The cell-cell interaction leads to a multistep events in in vivo situations. For example, hormone stimulation of fibroblasts is responsible for the release of surfactant by the lung alveolar cells. Androgen binding to stomal cells stimulates the prostrate epithelium. In other words, hormones, nutritional factors and xenobiotics exert stimulating effects on the cells to function in a coordinated manner. Xenobiotics broadly refers to the unnatural, foreign, and synthetic chemicals such as pesticides, herbicides, refrigents, solvents and other organic compounds. It is impossible to study these cellular interactions that occur in the in vivo system with isolated cells or cells in culture. This has lead to the attempts to develop organ and histotypic culture with the aim of creating in vitro models comparable to the in vivo system.

The three types of such cultures are:

1. *Organ Culture:* In this type of culture, the whole organs or small fragments of the organs with their special and intrinsic properties intact are used in culture.
2. *Histotypic Culture:* The cell lines grown in three dimensional matrix to high density represent histotypic cultures.
3. *Organotypic Cultures:* A component of an organ is created by using cells from different lineages in proper ratio and spatial relationship under laboratory conditions.

ORGAN CULTURE

In the organ culture, the cells are integrated as a single unit which helps to retain the cell to cell interactions found in the native tissues or organs. Due to the preservation of structural integrity of the original tissue, the associated cells continue to exchange signals through cell adhesion or communications. Due to the lack of a vascular system in the organ culture, the nutrient supply and gas exchange of the cells become limited. In order to overcome this problem, the organ cultures are placed at the interface between the liquid and gaseous phases. Sometimes, the cells are exposed to high O_2 concentration which may also lead to oxygen induced toxicity. Due to the inadequate supply of the nutrients and oxygen, some degree of necrosis at the central part of the organ may occur. In general, the organ cultures donot grow except some amount of proliferation that may occur on the outer cell layers.

9

Hormones in Animals

The metabolic effects of growth hormone in muscle have been studied in normal and hypophysectomised rats, and, to a large extent in recent years, in the isolated diaphragm muscle. The effects observed have mainly to do with the carbohydrate metabolism of muscle, and these are not as yet clearly or immediately related to the growthpromoting actions of the hormone. The current picture of the effects of growth hormone on muscle is confused by the fact that, depending on the conditions of the experiments, the hormone sometimes seems to promote and sometimes to inhibit, the utilisation of glucose by muscle.

The inhibitory effects on occasion seem to be directly mediated by the hormone, but in certain other experiments the hormone appears to give rise to inhibitory substances which circulate in the plasma, so that the effect is indirect and takes time to develop. The details of much of this work have been well reviewed by several workers.

CHEMICAL INTEGRATORS OF HORMONES

Hormones are the chemical integrators of a multicellular existence, coordinating activities from daily maintenance to reproduction and development. The neurotransmitters released by axons are one class of chemical communicators that act on an adjacent cell, usually a muscle cell or another neuron. Hormones are a mostly distinct class of chemical communicators secreted by nerves, ordinary tissue, or special glands; they act on cells far removed from the site of their release. They can be proteins, single polypeptides, amines, or steroids or other lipids. Hormones travel to their place of action via the circulatory system and then match their particular configuration with a specific receptor molecule attached to a cell membrane or, more usually, located within the cell.

The nervous system coordinates the more rapid activities of animal life, such as movement, while the hormones integrate everything else. Only the larger, more complex animals, such as vertebrates and some arthropods, have special endocrine glands to produce hormones; other animals use nerve cells or tissues such as the gonads. Endocrine glands are another example of a partitioning of functions into separate organs, a system that increases efficiency but that requires a relatively large size to maintain. Greater specialization is also associated with greater difficulties in regenerating lost parts or preventing breakdowns in functions.

Although the list of hormones found in the mammalian body may seem large, the numbers are surprisingly low for the variety of functions they influence. Which of the multiple functions any one hormone regulates depends on the specificity of the receptors on or within cells. Because all hormones bathe all cells as a result of their transport by the circulatory system, it is more efficient to have a general messenger transported to a cell, where it elicits only one of many possible outcomes. As in the nervous system, the specificity of response lies in the organ that responds and not with the messenger that merely commands action. Chemicals that allow communication among individuals are called pheromones. Sexual attractants are the most common, but there are many other kinds.

COMPLEX PHENOMENON

The consensus of opinion, expressed in the preceding chapters, seems to have been that "growth is a very complex phenomenon. It is true that these creatures appear, even when studied with the best optical or electronic equipment, as rather simple structures; and, compared with higher plants and animals, they certainly do not seem to be as highly differentiated. Nevertheless, they display an astonishing diversity of functions which in many respects they share with the trees, oysters, butterflies, and elephants. Generally they are found swimming or floating about in an environment containing a more or less abundant variety of chemical compounds. From these they pick up or absorb special ones which as a result of metabolic activities are subsequently converted into cellular constituents. When a large enough quantity and variety of the latter have been produced, and when the individual organism has sufficiently increased in size, it divides into halves, each one of which then repeats the process. This continues until the food supply, or specific ingredients thereof, gives out, or until inhibitory metabolic products begin to interfere with normal development.

The amazing feature of growth, in a microbial culture as in living beings in general, lies in the fact that it involves the manufacture and orderly arrangement of the manifold structural and functional components which characterise the organism in such a way as to duplicate accurately an existing pattern. How this is accomplished is one of the major biological problems, fundamentally as complex in the lower as in the higher forms of life.

On the other hand, the comparative ease of handling microorganisms, their rapid growth, and the opportunity for rigourously controlling various environmental factors all combine to render them favourable material for experimental studies of the basic mechanism of the growth process. In several instances the successful resolution of a complex phenomenon into its constituent components, the latter more readily amenable to scientific analysis than the overall event, has resulted from kinetic studies. By relating the rate of the process to a number of variables, preferably one at a time, regularities and irregularities are often disclosed which serve as useful indicators of component steps. It therefore seems appropriate to examine what kinetic studies on the growth of microorganisms have contributed and may contribute to a deeper penetration into the problem of growth in general. Nearly all that is known about the kinetics of growth of microorganisms has been learned from studies of so-called growth curves.

By inoculating a specified culture medium with a suitable organism and determining, at the start and after various time intervals, the number of individuals present—frequently restricted to a determination of the number of viable cells—the requisite data are collected for constructing a graph which represents the relationship between the microbial population (or density) and the time elapsed since the initiation of the culture. The most convenient curves are those in which the number of individuals per unit volume is plotted on a logarithmic scale against time. Figure demonstrates the commonly encountered shape of such growth curves. Much attention has in the past been paid to an interpretation of the various phases of these curves. It will be clear that the first phase denotes a period in which a gradual transition takes place from a condition where an increase in numbers does not occur to one where the increase is exponential phase. This second phase is easily accounted for; it means that during this period all organisms divide regularly, with the daughter cells behaving in an identical fashion. Towards the end of this phase the slope changes back to zero, evidence that when this point is attained either no further divisions take place, or that the increase in cell numbers is exactly balanced by the disappearance or death of individuals. The transition period, often brief and rather sharp, would imply that:

- Either fewer and fewer cells are still in a position to divide, or
- That the division rate of the individuals becomes progressively less.

During the final phase the death of the cells is predominant; from the point of view of studies on growth this phase is of no interest. A considerable amount of time has been devoted to the first and originally quite enigmatic phase, referred to as the latency and lag period. As a result of various investigations, mostly before 1930, the attitude towards the problems it presents can be summed up as follows. In many cases the cause for the existence of a latency and lag period resides in the cells used as inoculum. It can be completely eliminated by inoculating from a culture which itself is in the exponential phase. This has been demonstrated by Chesney for bacterial, and by Phelps for protozoan cultures. The careful observations of Henrici have shown that the morphology of a bacterial species differs markedly according to the phase of the culture. It is therefore clear that the initial lag must be ascribed to a transformation of aged cells into "embryonic forms". Depending upon the age of the culture from which the inoculum is prepared, this change may require anywhere from minutes to many hours.

Similarly, a period of preliminary adjustment generally follows upon a change of environmental conditions, such as composition of the culture medium, change in temperature of incubation, etc. In that event a lag phase can be observed even if the new cultures are inoculated from a culture in the exponential phase, and the length of the lag depends upon the extent and nature of the change in conditions.

Growth curves of microorganisms are as a rule accurately reproducible. Chiefly responsible for this feature are:

- The rigourous control which the experimenter can exercise over the environmental conditions,
- The small size of the organisms.

The small size eliminates many complications connected with the accessibility of foodstuffs. In higher plants and animals transport of nutrients, often over considerable distances, may be a controlling factor for growth; in microorganisms this is not the case. Furthermore, the small size insures that large numbers of individuals are always involved. Hence a statistical treatment of results becomes superfluous; every experiment automatically yields data in which individual variations are represented. This will be obvious if it be realised that with bacterial cultures populations of about 10^9 organisms per ml. are the rule, and that even for pure cultures of protozoa numbers up to 10^6 per ml. are far from rare.

That not all individuals in a culture behave in an identical manner, and that the results obtained really express statistical averages, follow from the fact that growth curves of microorganisms are always smooth even if cultures are started with a single individual, where one might expect a discontinuity such as was first observed by Ellis and Delbrück in their studies on bacteriophage multiplication. More direct evidence for the existence of variation among individuals in bacterial cultures can be found in the investigations of Vaas and Doudoroff. In spite of these advantages, the kinetic approach has not yet been used very effectively or extensively. It is probable that this must chiefly be ascribed to the unwieldy and laborious methodology that used to be necessary for the determination of a growth curve, requiring either many sets of culture plates, or the actual counting under the microscope of thousands of cells for each point on the curve. The newer methodology of determining population densities by turbidity measurements with the aid of photoelectric apparatus, though not always entirely free from objections, may well bring about a renewed interest in this line of work, as is foreshadowed by the important studies of

Monod during the last few years. Kinetic investigations on cultures of microorganisms are eminently suited for establishing relations between growth and environmental factors, especially the nature and amount of nutrients. Based upon the concept developed by Liebig about a century ago, it could be anticipated that in microbial cultures growth would be limited by those ingredients which are present in "minimum" quantity. Experiments such as those of Meyer on the growth of *Aspergillus niger* in media with graded concentrations of phosphate and ammonium sulfate have gone far towards establishing the validity of Liebig's "law of the minimum" for microorganisms. But these were not, strictly speaking, kinetic studies; the measurements dealt with total crops as dependent upon nutrient concentrations rather than with rates of growth. Furthermore, these researches were carried out at a time when little was known concerning the exact mineral requirements of the organism used, which makes the interpretation of the results somewhat difficult.

Decidedly more conclusive in this respect have been the recent studies in connection with growth factor requirements of microorganisms. The mere fact that microbiological assay methods have been developed for many of the known vitamins and amino acids provides convincing evidence for the thesis that a single nutrient factor may determine the extent of growth. However, with a very few exceptions, these contributions too are based upon measuring final yields rather than rates. And where the latter are employed, as for example in the thiamin assay by means of yeast fermentation, it is the metabolic, not the growth rates, that are the basis for the determination.

Before 1942 only a few investigations on the kinetics of microbial growth in relation to nutrient concentration were reported. The results have been well summarised by Rahn in the following statement: With increasing amounts of food, the crop, *i.e.* the final number of cells in a culture, usually increases. With a liberal supply of building material, increase of energy food will cause no acceleration in growth, but a longer growth period, and, therefore, a larger crop. This is limited by accumulation of fermentation products or of inhibiting cell secretions. "With all food components increasing, the crop will still be limited by fermentation products or cell excretions. The law of diminishing returns becomes quite evident." While these conclusions were based upon experiments with bacteria and yeasts, they apply equally to the growth of protozoa in pure culture, as was first established by Phelps. His studies also revealed a direct proportionality between cell yield and food concentration within very wide limits, a situation which, wrote Phelps, "has not yet been demonstrated in the bacteria nor in most of the work on yeasts." Two years later, however, Dagley and Hinshelwood supplied the experimental evidence that in bacterial cultures also the rate of growth is independent of the concentration of nutrient materials over a wide range, while the final crop of organisms is directly proportional to this concentration.

The latter part of these conclusions was corroborated by studies on the growth of *Escherichia coli* in beef broth. All of the earlier studies on the kinetics of growth in relation to nutrient concentration were carried out with complex media in which not one but all constituents were present in varying amounts. It was only recently that Monod started his investigations in a manner more likely to furnish basic information. By using bacteria which can develop in a strictly defined medium, it was possible to determine the effect of variations in the concentration of a single nutrient constituent upon both the rate of growth and the final crop. So far, this approach has been restricted to experiments with sugars and related compounds, applied over a wide range of concentrations. With three different bacterial species it was established that, for a given carbon source, the rate of growth is independent of the substrate concentration, and the total yield is strictly proportional to it, down to such low levels as 20 mg. substrate per litre.

There are on record a number of observations which suggest a different response of microorganisms to the concentration of nutrients. Thorne, for example, claims to have shown that the rate of yeast growth in sugar media falls uniformly with increasing concentration of the yeast, a behaviour entirely at variance with the above-mentioned results. And Porter has recently resuscitated the concept of "biological space" as earlier developed by Bail. The argument here is that, because the total crop of bacteria in liquid cultures reaches a certain maximum regardless of the concentration of foodstuffs, growth must be limited by a space factor rather than by nutrients. The following passage from Porter book is worth quoting in this connection: "There is considerable evidence in favour of some of Bail's claims concerning space theory and *M* (maximum) concentration, but many of his points are not accepted at this time. Further work will have to be done on this subject—why bacteria stop dividing at a maximum speed when a certain population is reached—before any definite conclusions can be reached."

Most, if not all, of these aberrant results can be readily ascribed to the experimental conditions imposing a limiting factor which, in spite of repeated exhortations, is still insufficiently taken into account. This limiting factor is oxygen. The supply of this gas is governed by its very low solubility coefficient and unless special measures are adopted to insure a greatly increased rate of diffusion (*e.g.* by forced aeration, continuous shaking, etc.), its availability in liquid media is apt to fall below that required for an optimum growth rate of the organisms. The studies of Phelps, Rottier, Rahn and Richardson, and Anderson attest to the correctness of this statement. It seems more than likely that the "biological space" referred to above is simply a measure of the oxygen supply. Particularly the close packing of bacteria in colonies provides a convincing argument for the contention that space *per se* cannot be considered seriously as an important factor in growth. Thus the best accessible evidence indicates that the growth rate is not affected by the concentration of nutrients between very wide limits, and that the total yield, over an equally wide range, is proportional to the amount of food. This situation carries important implications which will next be considered. Let me reiterate that growth consists in part of the transformation of foodstuffs into cellular constituents. This, in turn, depends upon metabolism. Thimann has presented a number of cases in which growth responses are strictly paralleled by metabolic responses following upon the exposure of plants to auxin solutions. It is therefore of the greatest significance that the rate of metabolism, as has been repeatedly ascertained, is generally unaffected by concentration differences of the substrate down to extremely low levels. And it is not surprising that growth would follow the same pattern.

In many respects the relation between metabolism and growth is quite obvious. "No growth without metabolism" was a dictum as familiar since the latter part of the nineteenth century as the phrase "No thought without phosphorus" had been earlier. The reason for this has become apparent as the concepts of thermodynamics were developed. The growing cell must synthesise cellular constituents, and for these syntheses energy is required. As a first approximation the relation between growth and metabolism could thus be understood as one of energy linkage; the metabolic activities of the cell provide the energy necessary for the performance of synthetic reactions. Based upon this concept, various investigations have been conducted aimed at measuring the "energetic yield" of growth processes. By determinations of the amount of cell material that could be synthesised from a known quantity of substrate, of the caloric value of the cell material, and of the energy liberated by oxidations accompanying the observed growth, data were assembled which permitted the computation of such "energetic yields." In subsequent studies of the same general nature more precise relations were calculated by distinguishing between a "metabolism for upkeep" and a "metabolism for growth".

However useful and important this sort of information may have been, recent trends have brought about a considerable change in the primary approach to the problems of syntheses and growth. Biochemists in particular have come to recognise that the thermodynamic formulations beg the question which has become paramount in today's biochemistry, *viz.* that of the mechanism whereby the result is achieved. Applied to the problem under discussion, this means that the concept of an energetic coupling between synthesis and breakdown, no matter how sound fundamentally, is too vague to be permanently satisfactory, and that ultimately our task is to discover the mechanisms whereby such coupling is accomplished.

The most fruitful hypothesis that has been advanced in this connection assumes the existence of material links between breakdown and synthesis. It may, somewhat primitively perhaps, be paraphrased in this manner: instead of regarding biological syntheses as very special processes which, because they require energy, are basically different from the breakdown reactions in which energy is liberated, it would be more appropriate to describe syntheses as series of enzyme-controlled step reactions, each step comparable in every way but one with those which collectively compose the breakdown. The differentiating feature is that, whereas the breakdown reactions can always start with any one of the utilisable substrates in the food, the synthetic processes require particular substances which are not usually provided but which arise from the initial substrates as] intermediate products in the normal course of their metabolic degradation. Such products are characterised by being more unstable, *i.e.* more reactive, than their ultimate precursors, the foodstuffs proper, thus implying that their energy level has been raised with respect to the initial materials.

This concept, first enunciated by Kluyver some fifteen years ago, does not conflict with the earlier one of energetic coupling. Its great value lies in the penetrating insistence upon the functioning of an intelligible chemical mechanism which is capable of accounting for the transfer of energy. During the past few years tremendous strides have been made in this direction. Ranking high among these is the discovery of the phosphorylated carbohydrate now known as "Cori-ester," a compound which can serve as the immediate raw material for the spontaneous enzymatic synthesis of a number of di-and polysaccharides. No less significant are the studies of Lipmann on the labile intermediary metabolic products characterised by a high-energy phosphate bond. It has now been shown that these substances, too, can be used for performing a number of fundamental biochemical syntheses with isolated enzyme systems.

A special corollary of Kluyver's concept is that the molecular structure of the substrate and of the intermediate products to which this can lead during its breakdown, rather than the amount of energy liberated during this process, becomes the all important determinant of its potentialities as the starting point for the synthesis of cell materials. The investigations by Clifton and Logan, Doudoroff, and Bernstein on the assimilation of chemically related compounds at different energy levels provided suggestive results in this connection. Better yet can this situation be illustrated by recent studies on the synthesis of some characteristic cell constituents.

Having recognised a number of vitamins as parts of enzyme systems, it has become increasingly clear that the biosynthesis of such entities proceeds in a chemically intelligible manner, even though many of the steps involved in their manufacture may still be obscure. The mere fact that certain microorganisms are unable to develop in simple media without being supplied with preformed parts of enzymes shows conclusively that the supply of energy is not the governing factor, and that the availability of appropriate building blocks must be reckoned among the requirements for growth.

Such phenomena are now generally interpreted to mean that the cells lack the machinery for the synthesis of essential compounds, which must consequently be supplied. It has frequently been shown that the same organisms can also develop in an environment which contains, instead of the enzyme part itself, substances that are chemically more or less closely related to it, and from which the cellular constituent in question is elaborated. Corresponding observations have been made pertaining to the synthesis of specific amino acids, as the epoch-making investigations of Beadle and Tatum and their collaborators with "biochemical mutants" have so amply demonstrated. Through studies of this kind it has been possible to elucidate various steps in the biosynthesis of a number of cell materials and to develop the thesis that each step is controlled by a specific gene. It is impossible to do this field justice here; reference is made to some recent reviews. No student of the chemistry and physiology of growth is likely to deny the importance of these advances in our knowledge of biosynthetic mechanisms. Nevertheless the inclusion of this phase may appear as a digression from my topic. What have such problems to do with the kinetics of growth of microorganisms?

It will be evident that even the growth of a microbe is subjected to the "principle of limiting factors." If the synthesis of all cell constituents can proceed faster in one medium than in another, growth will obviously be faster in the former. Similarly, the rate at which a single component is being synthesised may, under special circumstances, be the determining factor for the rate of growth of the organism. In that case our understanding of biosynthetic mechanisms might be advanced by kinetic studies on growth, as well as by the more strictly chemical ones which hitherto have been the sole method applied.

And through a kinetic approach it may be possible to learn something about the synthesis of these cell constituents by organisms whose external environment does not need to contain a supply of closely related substances because the biosyntheses can proceed from a variety of simple ingredients of the medium. While no investigations have yet been conducted with this purpose in mind, some data are available which support the idea here advanced. For example, Monod includes in his monograph figures on the rate of growth of *Bacillus subtilis* in a complex and in a synthetic medium, the latter composed of inorganic substances and sucrose. Growth proceeds faster in the former than in the latter medium; the generation times are 36 and 50 minutes respectively. Such differences could well be due to differences in the rate at which particular cell constituents become available, especially in view of the fact that in the synthetic solution they must all be built up from inorganic ingredients and sucrose.

If it were possible to increase the growth rate in the simple medium through the addition of known chemical entities, it might also become clear what special syntheses are responsible for the lower growth rate. Another instance is provided by Whelton and Doudoroff study on the assimilation of sugars and related compounds by *Pseudomonas saccharophila*. It includes a determination of the generation times of this bacterium in a standard mineral medium with any one of a number of carbon compounds. These were found to vary from 105-110 minutes in the presence of trehalose and sucrose, to 196 minutes in the presence of acetate; cultures in glucose and lactate exhibited an intermediate behaviour with generation times of 178 and 136 minutes respectively.

Studies of this kind, if sufficiently extended, could yield information concerning the nature of intermediate products which serve as building blocks and the rate at which these are generated. Correlation of such data with overall growth rates, supplemented with experiments on the influence of special additions selected for the purpose of furnishing similar or different building blocks, might then reveal the nature of limiting synthetic reactions. The above-suggestion approach is an easy and certain way of solving the problem of growth. They are intended to emphasise that kinetic studies of the growth of microorganisms, hitherto rather neglected, should be seriously considered as potentially capable of supplying information which can contribute towards a more detailed analysis of the manifold and interlinking events which, in their totality, constitute the growth process. The foregoing reflections on the effects of external factors on the growth rate of microorganisms, and on the interpretations and deductions that can be derived from the observed results, may have fostered the impression that internal factors can be regarded as more or less constant. But this is not the case; there exists an impressive body of indisputable evidence to the contrary. Changes in the morphological, physiological, and biochemical behaviour of microbial cultures, attributed to adaptations, modifications, mutations, etc., have long been recognised as manifestations of the intrinsic variability of the organisms.

Monod's recent studies on the growth rate of bacteria in media containing two different carbon sources instead of a single one offer an important advance in this field. Depending upon the nature of the organism and of the substrates, either of two types of growth curves describes the observed response of the bacteria. One represents a single period of exponential growth extending over the entire time interval during which substrate is available, with the sharp break at the end marking the complete utilisation of the two carbon sources. The other kind consists of two distinct parts, each in itself a characteristic growth curve, with its phases of accelerated, maximal, and retarded growth rates, the two curves being conspicuously separated by a horizontal stretch which indicates the cessation of growth during the corresponding period.

Monod has designated this latter behaviour by the term "phénomène de diauxie" ("diphasic growth") and has shown that it involves the occurrence of an adaptation process. By chemical analyses of the medium at different times it was ascertained that the first part of the growth curve corresponds to growth of the organism at the expense of only one of the two substrates, to the complete exclusion of the other. The second, separate growth curve coincides with the decomposition of the second substrate. And the intervening period during which the population remains constant represents the interval needed by the bacteria to adapt themselves to the utilisation of this second substance. Adaptations of this sort have been known for several years and have been ascribed to an adaptive enzyme formation.

They are strictly dependent upon the presence of a specific substrate, an observation which has led Yudkin to develop an ingenious hypothesis concerning the mechanism by which adaptive enzymes are generated, based upon differences in stability of the free enzyme molecules and of the enzyme-substrate complex. Monod's experiments have shown that the rate of such adaptations in bacterial cultures depends not only upon the presence but also upon the concentration of the substrate. In view of the well-established fact that metabolic rates are generally independent of this last-named factor, and also that the growth rate of the adapted bacteria is constant over a wide concentration range, Yudkin's hypothesis may have to be modified.

HORMONE-DEPENDENT SEX DIFFERENCES IN GROWTH

Hormone-dependent sex differences in growth rate have been known for a long time. It has also been known that growth rate and FCE (feed conversion efficiency) are higher in intact males than in castrates. It was natural, then, that the availability of hormones and other natural or synthetic substances displaying hormonal activity led to experiments aiming at their use to increase production.

Beginning in the mid-1950s, DES (diethylstilboestrol) and hexoestrol were administered to cattle increasingly in the US and the UK respectively, either as feed additives or as implants, and other types of substances also gradually became available. In general, such treatment has resulted in 10–15 per cent increases in daily gains, similar improvements in FCE and improvement of carcass quality (increased lean/fat ratio). Thus there has been a substantial reduction in the amount of energy required per unit weight of protein produced (1,2), and the economic implications of this have been great.

While the use of hormonally active substances in animal production rose, opposition to their use also increased, because of the theoretical possibility that residues in edible tissues might endanger consumers. The factors leading to the ban on DES in the US, first imposed in 1973, have been described (3). Several reports confirm that DES endangers the health of animals and man, when repeatedly used in large doses (4,5). However, as regards risks due to the presence of residues in meat produced according to regulations, no documented deleterious effects have ever been reported in man, either from DES or any other substance with hormonal activity.

A distinction should be made between the hormones as such, for which the metabolism in the body is relatively well known, and synthetic or other substances for whose metabolic inactivation the body may not possess the enzymes necessary. When natural hormones are used in animal production, claims of zero-tolerance residue levels are not meaningful, since these compounds occur in detectable and highly variable concentrations in body fluids as well as in the tissues of all animals, treated or not (6,7). For other substances with hormonal activity the situation is different. However, when residue levels are extremely low, it seems reasonable to weigh the potential risks against the undisputed positive effects some of these compounds have in animal protein production.

This chapter will discuss types of substances with hormonal activity currently in use or under investigation, their effects, mechanism of action, metabolism/elimination, tissue levels, risks to the consumer and their economic importance. Finally, other avenues to increased animal production as alternatives to use of hormones will be briefly envisaged. For the sake of simplicity the term hormone will be used, even if incorrectly, to cover all substances with hormonal activity, whether natural or synthetic. Since much information on the question collected before 1975 has been reviewed previously (8), the main emphasis will be placed here on research since that time.

HORMONE PREPARATIONS USED IN ANIMAL PRODUCTION

Hormones of endogenous origin

These comprise the "classical" steroid sex hormones, oestradiol-17β, testosterone and progesterone. The two former are used either in the free form or as esters, mainly those of propionic or benzoic acid. Esterification generally causes prolongation of the half-life of the compounds in the body by 40 to 50 per cent. The natural hormones having low bioavailability when administered orally, owing to rapid conjugation and metabolic transformation in the liver, they are therefore administered by subcutaneous implantation.

Hormones of exogenous origin

Of the oestrogens, the stilbene derivatives diethylstilboestrol (DES) and hexoestrol possess high biological activity and have been used most widely. They are active orally as well as by implantation. Other orally active oestrogens include ethynyl-oestradiol, a more slowly metabolized derivative of the true hormone, with higher activity. An oestrogen with an entirely different structure is zeranol, a derivative of a resorcylic acid lactone occurring in the fungus Giberella zeae. The synthetic androgens comprise a large number of substances, most of which are steroids. Of these, trenbolone acetate (TBA) possesses strong anabolic properties and has received much attention during recent years, used alone or in combination with an oestrogen. Another anabolic steroid is methyl-testosterone. Of synthetic gestagens, only one will be mentioned here: melengestrol acetate, which stimulates growth in heifers but not in steers, and which can also be used for the suppression of oestrus. Numerous other gestagens also exist, but at present few other than progesterone and melengestrol acetate are used to stimulate growth. In addition to these substances, numerous others exist, and some of them are used more or less frequently in clinical veterinary medicine. However, clinical applications of hormones are not considered to be of consequence to the consumer, since such treatment is much less frequent than the use of hormones to promote growth.

Hormone preparationsin current use as growth stimulants, which also shows modes of application, dosages, etc. It will be noted that almost all preparations currently in use are based on implantation, the site usually being the base of the ear, or less frequently, the dewlap.

RANGE OF APPLICATION

In cattle the use of hormones is limited to veal calves and beef cattle. Veal calves are produced mainly in continental Europe, to an extent of about 8 million per year. Research has demonstrated that hormone treatment improves growth rate, nitrogen retention and FCE during the five- to six-week period before slaughter. Beef cattle, including steers as well as heifers, were treated in large numbers, especially in the USA and the UK, with DES or hexoestrol, administered orally, until the use of these compounds was restricted. During the last several years, practice has changed dramatically in the direction of increased use of implants of natural steroids, synthetic anabolic steroids and the phyto-oestrogen zeranol.

Poultry generally do not appear to respond to oestrogens by increased gain but by changes in lipid deposition. In male and female turkeys, androgens have recently been reported to increase growth rate as well as FCE.

MODES OF APPLICATION

When DES was used as a feed additive, a usual procedure was to start treatment of steers at a body weight of 360 kg and continue administration for 120 to 170 days. Since restrictions on its use were imposed, most preparations have been administered as implants, whose effect is usually limited to 80 to 100 days. Practice varies with management systems. Animals may be implanted at live weights from 270 to 450 kg. Depending upon the age and weight at the time of implantation, the animals are either slaughtered at the end of this first period, or fed for an additional period, either without further treatment or after a second implant to act for another 80 to 100 days. Most types of implants in use are not removable, but removable types have recently been tested and their effects described. When tested in steers, no reduction in performance was recorded when the implants were withdrawn 32 and 39 days before slaughter. Implantation is subcutaneous, usually at the base of the ear, thus eliminating the risk that residues of the implantation site will be present in edible tissue.

MECHANISM OF ACTION OF HORMONES

No reliable explanation of how the growth-promoting hormones act has yet been furnished. Some observations indicate an indirect influence through changes in the balance of endogenous hormones. Thus there have been reports of DES and TBA increasing the levels of growth hormone and/or of insulin in plasma ; these hormones are known to stimulate amino acid transport across the cell membrane. However, others have found no such effect. Bulls fed DES (10 mg/day) over two years had significantly higher plasma testosterone levels than controls ; those levels are positively correlated with growth. Recent experiments indicate that DES reduces the rate of muscle catabolism in steers. As regards the anabolic androgens, evidence exists indicating competition with glucocorticoids for receptor sites on the muscle cell membrane. Since glucocorticoids have a catabolic effect on tissues, their displacement from muscle cells would reduce catabolism. TBA alone, and even more when combined with oestradiol-17β, causes a marked decrease in the concentration of total thyroxin in plasma of steers. In another study, combined oestradiol-17β-progesterone implants (20 + 200 mg) in steers caused a uniform but slight increase in thyroxine binding capacity. The significance of these findings is not yet clear.

EFFECTS OF HORMONES

Veal calves

In veal calves, hormone treatment may begin at a body weight of about 65 kg, the animals being slaughtered at about 170 kg. Implants of 20 mg oestradiol-17β + 200 mg progesterone in males and 20 mg oestradiol-17β + 200 mg testosterone in females resulted in a 20 per cent increase in daily gain and 21 per cent higher nitrogen retention in the period studied. In other studies, improvements of 10 to 12 per cent in gain and 10 per cent in FCE have been reported. Nitrogen retention is about 70 per cent in the very young veal calf, but decreases gradually to below 40 per cent at the age of about 15 weeks. For ages of 10 to 15 weeks, the average conversion of feed protein to body protein is about 40 per cent; this rate can be increased to about 60 per cent by hormone treatment. The effective preparations were DES, oestradiol-17β, and the combination of TBA + oestradiol-17β. More recently, positive effects have been reported for zeranol alone (36 mg) and for zeranol (36 mg) + TBA (140 mg), with increases in nitrogen retention of the same order as for DES and E2 +

TBA. When zeranol + TBA was implanted at the age of 56 days, the growth rate up to day 106 increased by 18 per cent.

Steers

The most extensive studies of the effects of hormones on growth and FCE have been carried out on steers, under strictly controlled conditions as well as in the field. Since 1975, most studies have involved implants of oestrogens alone, androgens alone, or combined oestrogen/androgen preparations, although many trials have also been based on oestrogen/progesterone combinations during recent years.

Oestrogen implants have included DES, hexoestrol, oestradiol-17β and zeranol. DES implants have, as in previous studies, resulted in an increase of about 12 per cent in gain and in improvement in FCE of the order of 10 per cent. Hexoestrol implants, usually in doses of 30 to 60 mg, have been shown in numerous experiments to lead to considerable improvement in growth rate and FCE. In 19 trials carried out over the years on experimental husbandry farms in the UK, the overall average increase in gain produced by 45 or 60 mg hexoestrol implants was 0.16 kg a day, and in only 2 of the trials was it less than half that figure. Oestradiol - 17β implants alone (30 mg) have resulted in a 24 per cent increase in gain and a 13 per cent improvement in FCE. Zeranol implants, usually at 36 mg, have consistently improved gain as well as FCE. In a series of 21 UK trials over several years, the average response to zeranol implants alone was an increase in daily gain of 0.15 kg. Only in one trial was there no response. Similar results have been obtained in Ireland. Positive effects on gain in steers have been observed under a variety of experimental conditions, under controlled feeding, on ad lib feeding of standardized rations, and on pasture.

TBA implants administered alone at a dose of 300 mg have also had positive effects on growth, even if combination with an oestrogen has yielded better responses (vide infra). In a series of 8 trials in the UK, the average additional daily gain amounted to 0.09 kg, with considerable variation among trials. Similar results have been reported from Ireland.

Combined preparations. A number of trials have been carried out with implants containing two hormones. The combination of an oestrogen with an anabolic steroid, or with progesterone, has met with the greatest responses. Synovex-S has consistently increased gain as well as FCE, with responses averaging about 20 per cent and 17 per cent respectively. Hexoestrol + TBA (usually 30 or 45 mg hexoestrol + 300 mg TBA) has resulted in marked increases in gain, of the order of 30 per cent and in FCE of the order of 20 per cent. Oestradiol-17β + TBA (20/140 mg) has given similar results, as has Zeranol + TBA (36/300 mg), also recently tested. Hormone preparations have also been tested in combination with substances such as monensin, which increase FCE by promoting propionic acid formation in the rumen. Results have varied from no effect to marked additional gain. Reimplantation, tested under various forms of management with varying results, has not gained general acceptance. Lamming has stated that "repeat implantation of hormone is not likely to produce the benefits obtained from its initial use, since a second dramatic change in the endocrine balance of the animal is not likely to occur. In addition, double implantation increases the possibility of exceeding the optimum dose rate and the chance of deleterious side effects occurring." The evidence for highly significant positive effects on the growth rate and FCE of steers is thus beyond dispute, the most marked effects being provoked by implants combining an oestrogen with an androgen of high anabolic activity.

Bulls

Since the entire male animal produces its own anabolic androgen, testosterone, an effect of additional hormones similar to that for steers is not to be expected. The number of trials with bulls is also limited. Positive effects on gain have been reported using DES alone and combined oestrogen/TBA implants ; in other studies, no effect on gain has been recorded, while a certain increase in the deposition of fat in the carcass has been observed.

Heifers

Recent trials with beef-producing heifers have mostly been based on the use of an androgen, although oestrogens have been tested, alone or in combination. Thus, zeranol has been reported to increase gain, while in other trials no response has been observed. TBA administered alone (300 mg) has led to increases in weight gain and FCE of the order of 36 per cent and 25 per cent respectively. In other trials, combinations of an oestrogen with TBA or testosterone have yielded significant growth responses. In general, it appears that the effect of TBA alone in heifers corresponds closely to the effect of combined oestrogen/TBA implants in steers.

Sheep

Trials have mainly concerned wether lambs, and positive effects of hormonal treatment have been reported using DES, hexoestrol and zeranol, although other reports have indicated that zeranol yields no significant effects. Wether lambs implanted with TBA + oestradiol-17β have shown increases in gain, carcass weight and FCE. In general, however, the results obtained in sheep thus far do not warrant the same clear-cut conclusions as for steers and heifers.

Swine and poultry

There is little evidence that existing hormonal preparations influence the growth rate and FCE to an extent that would be interesting from a practical point of view. The lean/fat ratio in male castrate and female pig carcasses may be increased by the use of oestrogen/androgen combinations. In poultry, redistribution of fat in the body is a known effect of oestrogens. Recent research indicates improved growth rate and FCE using androgens in young male and female turkeys.

UNDESIRABLE SIDE EFFECTS IN TREATED ANIMALS

Reported side effects of hormone treatment for growth stimulation are few and generally concern the use of oestrogens in steers. Changes in body conformation such as feminization and raised tail-heads were described as early as 1958. Similarly, bulling has occurred with increased frequency, although in most animals it is limited to the first few days after implantation. However, it has been reported from Kansas that 2.2 per cent of all steers fed in pens have to be removed, at an estimated loss of $23 per head. In a study of the effect of reimplantation of oestrogens in steers, all animals were given a 30 mg DES implant at a live weight of 260 kg, and then reimplanted 91 days later, with either 30 mg DES or Synovex S. Following the second implant, the frequency of the steer-buller syndrome was 1.65 per cent for the DES-DES group, and 3.36 per cent for the DES-Synovex S group. The economic advantage of using DES + DES was estimated at $1.15 per head. The steer-buller syndrome is a special problem in feedlots.

METABOLISM, ROUTES AND RATES OF ELIMINATION

The general patterns of metabolism and elimination of endogenous hormones in farm animals have been outlined. In ruminants, testosterone and oestradiol-17β are rapidly converted to their epimers, biologically much less active, epitestosterone and oestradiol-17a. Progesterone is partially converted to androgens before excretion. In the pig, epimerization of testosterone and oestradiol-17β does not appear to take place to a significant degree. The faecal route of elimination dominates in ruminants, while in the pig urinary excretion is more important.

Progesterone

After repeated injections of progesterone to cows and steers over 2 to 3 weeks followed by 14C-progesterone for 2 to 5 days, the animals were slaughtered 2 to 3 hours after the last injections. Activity levels were 2 to 7 times higher in the fat, 3 times higher in the kidneys, and 13 times higher in the liver than in the muscle. Excretion of radioactivity amounted to 50 per cent and 12 per cent in faeces and 2.0 per cent and 1.2 per cent in urine in cows and steers respectively. About 50 per cent of the activity in muscle and milk was associated with unchanged progesterone, most of the remaining activity being associated with a mono-hydroxy compound. Cooking or frozen storage did not affect the nature or quantity of metabolites.

Oestradiol-17β

Following daily injections of 1 mg oestradiol-17β or its benzoate to heifers and steers for 11 days, followed by the 14C-compounds on days 12, 13 and 14, the animals were slaughtered 3 hours after the last injections, when residual levels were maximal. In muscle extracts, oestradiol-17β represented the major fraction of extracted activity (38 to 71 per cent), followed by oestrone (17 to 45 per cent). Levels in muscle were 161 to 225 pg/g and 40 to 86 pg/g for oestradiol-17β and oestrone respectively. In fat the levels were 3 to 5 times higher. The authors conclude that residual levels are extremely low when these hormones are administered as growth stimulants to growing/finishing cattle. Glucosides of the 17β- and the 17a- epimers, and the glucoronide of the 17a- epimer are the major metabolites in cattle. When oestradiol-17β was administered orally to swine, plasma concentrations were very high 7 min after administration. Oestradiol was completely conjugated during absorption and its first passage through the liver. Some conversion to oestrone took place.

DES

The metabolism of DES in food-producing animals has been reviewed recently. The substance seems to be eliminated to a large extent in unaltered form. After oral administration of 14C-DES to beef cattle, 99.5 per cent of the radioactivity was excreted within 5 days after withdrawal. In liver extracts, radioactivity associated with DES-conjugate and free DES was found to be 75 per cent and 25 per cent respectively. Higher than background levels of activity were observed after withdrawal in kidney, liver, bile and urine/faeces for up to 5, 7, 9 and 11 to 12 days respectively.

The fate of 24-mg DES implants containing 14C-DES and implanted in the dewlap of calves was studied over 98 days. Free radioactivity was almost completely associated with unchanged DES. At the time of slaughter, levels were less than 0.1 ppb in muscle and fat, and 1 to 1.5 ppb in liver and kidney. In a study in steers implanted with 14C-DES, on day 120 after implantation radioactivity in muscle was not distinguishable from background. It was above background in spleen, lung, adrenal glands and kidney, but less than levels corresponding to 0.5 ppb.

In a similar study on steers, 120 days after implantation, levels in liver, kidney, lungs and salivary glands were in the range of 0.07 to 0.13 ppb of DES equivalent. In a recent study of DES metabolism in rhesus monkeys and chimpanzees, most of the substance was excreted with the urine. Extracts in the organic and aqueous phase mostly contained unchanged DES in the free and conjugated form respectively. Current evidence indicates that the oxidative metabolism of DES leads to at least three compounds that may have cytotoxic or mutagenic activity, but these have not been identified as DES metabolites in ruminants, but in the mouse.

Trenbolone acetate (TBA)

Trenbolone is a 17β-OH steroid esterified in the 17 position with acetic acid. Upon release in the organism the ester is rapidly hydrolyzed to the free compound TB-17β-OH and acetate. In cattle the 17β-OH compound is rapidly transformed to its 17a-OH epimer, in the same manner as oestradiol-17β in this species. The 17a epimer possesses only about 5 per cent of the biological activity of the 17β epimer. Another metabolite of TBA in cattle is the 17-keto compound, analogous to oestrone; quantitatively it appears to be of very little importance. Following intravenous injection of TBA, levels of TB-17β-OH and TB-17a-OH of 0.05 and 0.005, 0.10 and 1.0, 0 and 191 ppb have been recorded for muscle, liver and bile respectively. Other metabolites occurred in extremely small quantities in cattle. Similar findings have been reported in studies based on the use of implants. The major route of excretion is by faeces. Metabolism studies of TBA thus clearly show that the substance is rapidly subjected to biological inactivation in cattle, mainly by epimerization of the free steroid to the 17a-compound, and that the major route of excretion is via the bile.

Zeranol

Using a gas chromatographic method with a sensitivity limit of 20 ppb, no residues of zeranol could be detected in edible tissue from cattle slaughtered 65 days following implantation of 36 mg, or from lambs 40 days following implantation of 12 mg. In another study, tritiated zeranol was implanted in cattle as part of 36-mg doses. Skeletal muscle obtained 10, 30 and 50 days following implantation contained no detectable residual activity. This confirms previous results based on the use of 14C-labelled zeranol.

HORMONES IN FOOD: MEAT FROM HORMONE-TREATED ANIMALS VERSUS OTHER SOURCES

According to the Agricultural Research Service, United States Department of Agriculture (ARS), the average per caput consumption of beef is 157 g per day in the US. Calculations show that 157 g of beef from an animal implanted 61 days before slaughter with a combined implant containing 20 mg oestradiol-17β + 200 mg progesterone or testosterone will contain 3.43 ng oestrogen and 19.5 ng progesterone or 16 ng testosterone. Table, which provides data on normal levels of these hormones in certain dairy foods, shows that some foods represent hormone sources vastly richer than meat from hormonetreated animals. Based on these values, and averages for consumption of various foods, the relative contribution of meat from hormone-treated animals to the total consumption of hormones has been calculated on the assumption of proper use of the hormones. It is clear that in most cases the contribution from meat of treated animals is insignificant when hormones have been properly used, and must be considered to be biologically without impact. This becomes even more evident when seen in relation to normal endogenous hormone production in man, as illustrated in Table. It will be seen that even for oestrogens, the hormones considered the greatest risk, the maximal contribution from meat (assuming proper use of the hormones) is less than 0.01 per cent in the prepubertal boy who represents the lowest endogenous oestrogen production.

RESIDUES IN EDIBLE TISSUES OF HORMONE-TREATED ANIMALS

Much work has been devoted to the development of sensitive methods of detecting hormone residues in meat from hormone-treated animals. As regards compounds given orally, it should in principle be possible to realize claims of zero-tolerance residue levels, by selecting the proper withdrawal time. During recent years, the use of implants has, however, gained in importance. While removable implants have been tested in steers, with no decrease in performance when withdrawn 32 and 39 days before slaughter, the wide use of non-removable implants makes residue studies important. Determination of normal levels of endogenously produced natural hormones is also important, to enable risk evaluation to be carried out in realistic terms.

Several residue studies have been made of synthetic as well as natural compounds, mainly in cattle. When regulations governing dose, sites of implantation and timing in relation to slaughter are adhered to, residue levels of DES, hexoestrol and oestradiol-17β in edible tissues have generally been in the lower ppb to the ppt range, *i.e.* from a few ng/g down to some hundred pg/g of tissue. In the latter case there was almost complete overlap between values for untreated and treated steers after 105 days. Zeranol implants have so far not left detectable residues in edible tissue.

Most studies of androgens have concentrated on TBA. The ester being rapidly hydrolyzed, measurements of residues have been limited to the free compound and/or its major metabolite. Results based on radio-immunoassay of extracts or on radioactivity measurements have indicated levels in edible tissue of the order of 1 ppb or below. In a recent study using implants containing tritiated TBA in heifers, it was found that when slaughter took place 60 days after implantation, the major proportion of tritium-containing residues was not extractable with organic solvents.

In muscle 95.5 per cent, in liver 94.4 per cent, in kidney 98.8 per cent and in fat 59.1 per cent of the radioactivity remained in the aqueous phase, not quantifiable by radio-immunoassay. This suggests that the major part of the residues after TBA implantation occurs in a non-extractable, possible covalently bound form in tissues. Residue levels of gestagens have been also measured, in connection with their use as growth stimulants. Residues of melengestrol acetate used as a feed additive in daily doses of 0.25 to 0.50 mg per head have consistently been below the sensitivity levels of the methods used (*i.e.*, below 10 ppb in fat, liver, muscle and kidney), whether or not the compound was withdrawn 48 hrs before slaughter.

CARBOHYDRATE METABOLISM IN HORMONES

One of the striking effects of hypophysectomy in the rat is that the glycogen of voluntary muscle disappears much more rapidly than normal during a fast, falling in 16 to 24 hours to about half the level of the fed state, which is usually not very different from that seen in normal fed rats. The cardiac glycogen, which rises during a fast in the normal rat, also falls to low levels in the fasting hypophysectomised rat.

Since there is rapid depletion of the liver glycogen in the fasting hypophysectomised rat, it might be thought that the failure to maintain the muscle glycogen could be attributed to the early exhaustion of the hepatic sugar store and to the diminution of gluconeogenesis in consequence of the adrenal atrophy after hypophysectomy. In fasting adrenalectomised rats, however, the muscle glycogen stores are well maintained and the cardiac glycogen increases normally, despite the rapid loss of liver glycogen and the diminution of gluconeogenesis characteristic of these animals. The respiratory quotient of the hypophysectomised rat remains elevated during a fast. If the disposition of a carbohydrate meal is studied in such animals, it is found 4 hours after the meal, that there is less liver and muscle glycogen and a lower blood sugar, and that a much greater proportion of the glucose absorbed appears to be oxidised. Hypophysectomised rats with the basal metabolic rate restored to normal levels with thyroxine exhibit the same abnormalities, but the absolute rate of utilisation of carbohydrate is still greater. In the eviscerated, functionally hepatectomised, hypophysectomised rat, the blood glucose and muscle glycogen fall more rapidly than normal.

The rate at which glucose must be infused to maintain the blood sugar constant is about 25-27 mg. per 100 g. per hour as compared to 11-13 mg. per 100 g. per hour in the eviscerated normal rat. Similar rapid rates of glucose utilisation were observed in hypophysectomised eviscerated rabbits by Greeley. When it is recalled that the hypophysectomised animal, with its low level of thyroid activity, has a much lower metabolic rate than the normal animal, these differences are all the more remarkable.

If fasting hypophysectomised rats are treated with alkaline extracts of anterior pituitary tissue (APE) or with purified growth hormone during the period of fasting, the fall in muscle glycogen is prevented and the cardiac glycogen rises rather than falls.

In fasting normal rats, the injection of growth hormone during the last 6 hours of a 24-hour fast also leads to a greater accumulation of glycogen in the gastrocnemius muscle, the diaphragm, and the heart. In these experiments, there were no significant changes in liver glycogen, so that the effects of growth hormone appear to be exerted chiefly in conserving muscle glycogen. Of the different muscles observed, the heart muscle appeared to be most sensitive to growth hormone, the diaphragm least sensitive.

The effect of growth hormone in conserving muscle glycogen in the fasting animal may be an example of a general tendency of the hormone to diminish carbohydrate utilisation. Another aspect of this tendency may be seen from its effects on the disposition of fed carbohydrate in the intact rat. Fasted rats given a carbohydrate meal respond, during the 4 hours after feeding, with a rise in R.Q., and an increase in liver and muscle glycogen and blood sugar.

If insulin is given with the meal, the rise in R.Q. is greater, there is a greater increase in muscle glycogen, and the liver glycogen and blood sugar are diminished. If APE or purified growth hormone is given prior to the meal, there is a still greater increase in muscle glycogen, the blood sugar and liver glycogen are about the same as in the normal rat, but the R.Q. does not rise nearly so high as after carbohydrate alone.

Finally, if both insulin and growth hormone are given with the meal, there is a spectacular accumulation of muscle glycogen, the liver glycogen and blood sugar are intermediate between those seen in the untreated and the insulintreated rats. In experiments in another design, Illingworth and Russell also showed that insulin and growth hormone together had greater effects than either hormone alone on the glycogen of gastrocnemius, heart, and diaphragm of fasted rats given a glucose meal. All of these observations indicate that growth hormone affects the disposal of available sugar in favour of storage in the muscle rather than oxidation or conversion. It is interesting to note that, although the two hormones have opposing effects on the blood glucose, insulin and growth hormone work together in promoting the storage of glycogen in muscle. Other aspects of the effect of growth hormone on carbohydrate metabolism are presented by the work of de Bodo and his colleagus on the hypophysectomised dog. The hypophysectomised dog in general behaves in a way quite similar to the hypophysectomised rat; the animals are highly sensitive to insulin, and, in the postabsorptive state, have much lower blood sugars than normal dogs. In experiments using C^{14}-glucose as a tracer, the interesting observation is made that in hypophysectomised dogs 16-18 hours after the last meal, the rate of uptake of glucose by the peripheral tissues and the rate of hepatic output of sugar are lower than normal.

The rate of the CO_2 production in these animals is also much less than normal, as one might expect in hypophysectomised animals with much diminished thyroid function; one wonders, therefore, what portion of the total metabolism is contributed by the blood glucose, and what might be seen if these observations were made with hypophysectomised dogs brought to a normal metabolic rate with thyroxine.

As they stand, however, these experiments give no evidence of an excessive continuing utilisation of plasma glucose in the postabsorptive period. The injection of a small dose of insulin is followed by a greater than normal increase in the rate of uptake of plasma glucose by the peripheral tissues. The ensuing hypoglycemia is relatively even more severe, however, because the liver (which still contains adequate glycogen) fails to respond adequately to the falling blood sugar level.

The exaggerated insulin sensitivity of these animals therefore has both a peripheral and a hepatic component. If hypophysectomised dogs are treated with growth hormone for 4-5 days, the level of the postabsorptive blood sugar rises, the rates of glucose uptake and inflow are restored to normal, the increase in glucose uptake induced by the same small dose of insulin used before treatment is reduced to normal limits, and this is now smoothly compensated by an adequate increase in hepatic output of glucose, an effect evidently brought about by growth hormone but as yet without explanation. In normal and adrenalectomised dogs also, treatment with growth hormone increases the level of the postabsorptive blood sugar and the rates of inflow and outflow of plasma glucose. In the conditions of these experiments, therefore, although growth hormone reduces the sensitivity to insulin, there is nevertheless an increased rate of flow of glucose from the liver to the peripheral tissues. An antagonism of growth hormone to insulin in the sense of a marked depression of glucose uptake by the peripheral tissues is not seen.

One does not of course know what is happening to the glucose after it leaves the plasma. Data on the specific activity of the expired CO_2 would be most interesting and enlightening in this respect. The picture of these events in the growth-hormone-treated dog is not inconsistent with the observations cited above on the glucose-fed rat treated with growth hormone or insulin and growth hormone. The novel feature of these observations is the effect of growth hormone on the hepatic output of glucose, but speculation on this point is outside the range of a chapter on muscle.

EFFECTS ON DIAPHRAGM AND HEART IN VITRO

In recent years, a great many observations have been made on the effect of growth hormone on carbohydrate metabolism in the isolated rat diaphragm. The results of different investigators have varied a great deal in detail, partly no doubt because of differences in technique, sources of tissue, previous history and strain of animal, and differences in the nature of the hormone preparations used. It has generally been observed that diaphragm from hypophysectomised rats takes up glucose from the medium at a higher than normal rate, although this effect seems to be dependent in part on the glucose concentration in the medium; at the higher concentrations it is definite and reproducible.

There is also general agreement, with one doubtful exception, that growth hormone added *in vitro* to diaphragm from normal or hypophysectomised animals does not depress glucose uptake or diminish the effect of added insulin in increasing glucose uptake or glycogen synthesis. Instead, many workers have noted that growth hormone added *in vitro* has itself an "insulin-like" action, increasing the rate of glucose uptake of diaphragm from hypophysectomised animals. An insulin-like action of growth hormone may also be observed in normal rat diaphragm, but it is dependent on the nature of the incubation medium. Russell reported that growth hormone added *in vitro* to diaphragm from glucose-fed normal rats diminished the rate of fall of glycogen during incubation, but many subsequent attempts to repeat these experiments have so far failed. The injection of growth hormone into the animal prior to taking the diaphragm is followed by an interesting series of events. If the tissue is taken within 3 hours, only an "insulin-like" effect on glucose uptake is seen, but thereafter, reaching a maximum in 18-24 hours, the glucose uptake and the response to insulin *in vitro* are much depressed. This delayed effect of injected growth hormone is enhanced by the injection of adrenal cortex extract.

The absence of a direct effect of growth hormone *in vitro* and the development of a depression in glucose uptake in diaphragm removed from the animal some time after the injection of the hormone led to the suggestion that an altered form of the hormone, or some product released into the circulation after its injection, might be responsible for the altered behaviour of the diaphragm towards glucose.

In pursuit of this suggestion, Bornstein and Park observed that the serum of alloxan diabetic rats contained a factor that inhibited the glucose uptake of normal diaphragm. The factor was absent from the serum of hypophysectomised alloxan diabetic rats, but it reappeared after the injection of growth hormone and adrenal cortex extract, but not if either hormone preparation was given singly. Their factor appeared to have the properties of a lipoprotein. Vallence-Owen and Lukens have made a similar set of observations on plasma from normal, depancreatised, hypophysectomised-depancreatised and adrenalectomised-depancreatised cats. The plasma insulin activity of all of the depancreatised animals, measured by the effect on glucose uptake of the rat diaphragm, fell to zero, and, in addition, there appeared in the plasma of the diabetic cats a factor which greatly inhibited the effect of added insulin (1000 microunits).

The factor disappeared after hypophysectomy or adrenalectomy, and it was not restored by growth hormone alone or by adrenal cortex hormone alone. Since the inhibitory effect of the plasma from diabetic cats was not destroyed by freezing and thawing, it appeared doubtful that it could be a lipoprotein. The relation of this factor to that of Bornstein and Park remains to be determined. Another interesting system in which an *in vitro* effect of growth hormone has been observed is the isolated perfused rat heart. This preparation has the advantages that it has a proper circulation to the cells, an active metabolism, and good sensitivity to both insulin and growth hormone.

The rate of glucose uptake by the isolated heart, about 15 mg. per gram dry weight of heart per hour, is depressed about 50 per cent by the addition of 0.1 gm. per millilitre of growth hormone to the perfusion medium. In the presence of 2 milliunits per millilitre of insulin, the rate of glucose uptake is 40 mg. per gram dry weight per hour; 0.1 ¼gm. of growth hormone per millilitre reduces this to 28-30 mg. per gram dry weight per hour. Studies with galactose indicated that growth hormone had the effect of diminishing the galactose space, whereas insulin has the effect of increasing it. An analysis of the time course of galactose penetration showed that growth hormone, in the presence or absence of insulin, approximately halves the numerical value of a derived coefficient which is taken as a measure of the activity of the carrier system responsible for the entry of glucose or galactose into the heart muscle cells. Insulin, on the other hand, in the presence or absence of growth hormone, increases the activity of the carrier system about fourfold.

Since the actions of the two hormones are independent in the sense that each exerts its proportionate effect on the carrier system in the presence or absence of the other, growth hormone is not regarded as an "anti-insulin" but as a factor diminishing the scale of activity of the carrier system on which insulin acts. The authors point out that the observed effects of growth hormone and insulin on galactose penetration into the heart are entirely consistent with the observed relationship between the adenohypophysis and the pancreas: "First, the depressing effect of factor does not involve insulin; it occurs in its absence. Secondly, the effect of insulin is not dependent on antagonizing factor: insulin is more effective in its absence. There is, in fact, an *in vitro* Houssay phenomenon in our system."

Bronk and Fisher use the term "factor" because the growth hormone preparations used had more than one effect. The less pure preparation used had a complex, biphasic dose-response curve, in contrast to the more highly purified preparation derived from it. Both preparations became more effective on storage in dilute solution at-20°C. for several days, so that some labile factor acting in the same direction as insulin may also have been present. "Factor" *may* be growth hormone, but a decision on this point is quite properly reserved. This interesting work on an active, intact muscle preparation needs confirmation and extension, but it shows great promise of providing a simplified system in which the analysis of hormonal actions may be carried out under more favourable, more nearly physiological conditions than *in vitro* experiments with thin muscle strips can permit.

The picture of the action of growth hormone on carbohydrate metabolism in muscle is as yet by no means clear. There is evidence that it acts to conserve the carbohydrate stores and that it influences the disposition of carbohydrate within the muscle. There are indications that, in the presence of insulin, the entry of sugar into the muscle may be facilitated (the "insulin-like" effect in the diaphragm), or that (as in the heart) it may act directly to limit the activity of the carrier system that is facilitated by insulin.

The hormone also may give rise to substances which can depress the uptake of glucose by the muscle. In the intact glucose-fed rat, the hormone depresses the R.Q. and spares the oxidation or conversion of glucose in the tissues. Whether this is a direct effect, or a consequence of the mobilisation of an alternative substrate, fat, cannot yet be decided. There are many points of conflict apparent in the present observations, and it is not now possible to resolve them in terms of a single principle. It is evident that a great deal more work needs to be done, in a wider variety of conditions, and certainly on a wider variety of muscles, before a coherent picture of the action of growth hormone on muscle can be drawn.

GENERAL EFFECTS OF THYROXINE

That the muscles share in the general effects of thyroxine on the metabolic rate is well established. In the surviving tissues of hypo-or hyperthyroid animals, the muscles and heart have exhibited alterations in oxygen uptake of the same order as those seen in the intact animal, the cardiac tissue showing rather greater changes, skeletal and smooth muscle somewhat less. It is probable that the heart undergoes *in vivo* even greater alterations in metabolic rate than it shows *in vitro*; for in addition to the change in its own "basal" rate of heat production, the work required of the heart, in providing for the altered needs of the rest of the body, must be correspondingly affected. There has been no convincing demonstration that the thyroid hormone affects the metabolism or function of muscle in the mature animal in any other way than through its control of the metabolic rate and consequent effects. The mechanism of action of thyroid hormone on the metabolic rate has not been established.

Metabolism of Foodstuffs

The metabolism of carbohydrate and fat in muscle seems to undergo no special alteration with deficiency or excess of thyroxine other than those expected from the relationship of the supplies of foodstuffs to the demands for energy. The rates of catabolism of administered carbohydrates and ketone bodies and of body fat are affected about in proportion to the metabolic rate, whether it is high or low, and the proportionate disposition of fed carbohydrate in muscle is normal in hypothyroidism.

In hyperthyroidism, the glycogen stores of muscle may be depleted, especially in the heart; but the fall in the cardiac glycogen has appeared to be related quite closely to the work of the heart as indicated by the pulse rate, and similar effects on glycogen have been produced by other means of altering the heart rate. It is probable that similar relations of supply and demand obtain in skeletal muscle and liver as well.

Protein Metabolism

The relationship of the thyroid hormone to protein metabolism is biphasic. On the one hand, as mentioned earlier, small amounts of thyroxine are necessary for normal growth and development, so that in this sense the hormone is anabolic. On the other hand, the "basal" rate of protein catabolism is correlated directly with the basal metabolic rate. This has been seen in the nitrogen excretion during fasting or in subjects on nitrogen-free diets (the "endogenous" nitrogen catabolism); in the rate of release of amino acids from peripheral tissues of eviscerated animals; and in the release of protein and amino acids into the medium from isolated diaphragm.

When fasting thyroidectomised rats were given N^{15}-glycine, a larger proportion of the isotope was excreted and this was diluted to a lesser extent by endogenous unlabeled nitrogen. Hence, both the anabolism and catabolism of body protein appeared to have been diminished. Similar observations in patients with myxedema before and after treatment with thyroid hormone also indicated that the rates of synthesis and breakdown of body protein were both affected in parallel.

The rate of weight loss in denervated muscle was slower than normal in thyroidectomised animals, but occurred at the normal rate when the animals were given thyroxine in amounts which restored the normal metabolic rate. The effects of thyroid hormone on the net rate of protein catabolism evidently did not require innervation and were not dependent on muscle work. In both clinical and experimental hyperthyroidism, a negative nitrogen balance is frequently seen; and if the condition is prolonged, a considerable degree of wasting of the muscles and other tissues can occur. Since the negative nitrogen balance can be overcome by the provision of sufficient food, it is probably mainly the result of imbalance between supplies of foodstuffs and the increased metabolic rate. One may suppose that if anabolism of protein is increased in this state, it is augmented to a lesser degree than is the rate of catabolism. An increase in the requirements for and secretion of adrenocortical hormones, which has been postulated to occur in hyperthyroidism, might be expected to enhance the tendency to lose nitrogen in this condition.

Creatine Metabolism

Because of the weakness of the muscles characteristic of hyperthyroidism, much attention has been devoted to creatine metabolism in relation to the thyroid hormone. The subject has been reviewed extensively by Wang. Creatinuria, diminished creatine tolerance, and slightly lower excretion of creatinine are usual in both clinical and experimental hyperthyroidism. The total creatine content of the muscles, particularly of the heart, has been reduced in thyrotoxic animals and the phosphocreatine content also is said to be low. In skeletal muscle, the reduction in phosphocreatine was the same as that in total creatine in absolute terms, but larger proportionately; but the changes were rather small in any case. It may be remarked that in nearly all of the work of this type done in animals, the dosages of thyroid hormone used have been massive in the extreme—*i.e.*, just sublethal or sometimes even fatal in amount.

The changes so induced in creatine and phosphocreatine and other energy sources may be more expressive of non-specific damage than they are of thyroid hormone activity on muscle metabolism *per se*. From this work, no special alterations in creatine metabolism or in the functional aspects of creatine in muscle seem to have been uncovered. Since creatinuria and loss of creatine from muscle are seen in a variety of conditions in which atrophy or wasting of the musculature is prominent, it is probable that in hyperthyroidism also these features are related mainly to the loss of substance from the muscles. The relative depletion of phosphocreatine and other energy sources might be expected under conditions of relative overwork, as in the thyrotoxic heart.

10

Bacterial Pathogenicity in Animals

Some bacteria are parasites of plants or animals, meaning that they grow at the expense of their eucaryotic host and may damage, harm, or even kill it in the process. Such bacteria that cause disease in plants or animals are pathogens. Human diseases caused by bacterial pathogens include tuberculosis, whooping cough, diphtheria, tetanus, gonorrhea, syphilis, pneumonia, cholera and typhoid fever, to name a few. The bacteria that cause these diseases have special structural or biochemical properties that determine their virulence or pathogenicity.

These include:

- Ability to colonise and invade their host;
- Ability to resist or withstand the antibacterial defences of the host;
- Ability to produce various toxic substances that damage the host.

Plant diseases, likewise, may be caused by bacterial pathogens. More than 200 species of bacteria are associated with plant diseases, but a very small handful of genera are involved.

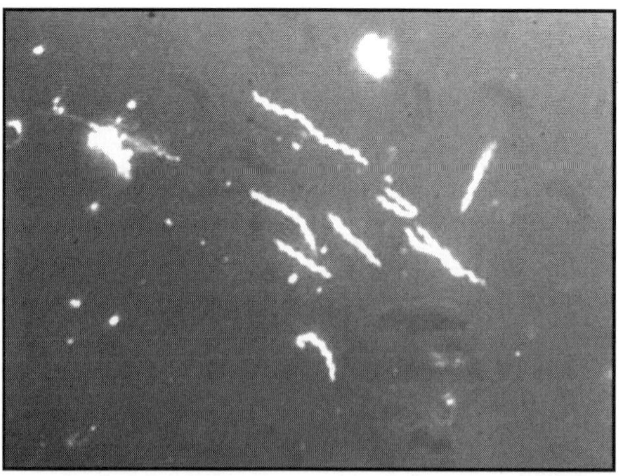

Fig. *Borrelia Burgdorferi.* This Spirochete is the Bacterial Parasite that Causes Lyme Disease. CDC.

EXPLOITATION OF BACTERIA BY HUMANS

Bacteria are used in industry in a number of ways that generally exploit their natural metabolic capabilities. They are used in manufacture of foods and production of antibiotics, probiotics, drugs, vaccines, starter cultures, insecticides, enzymes, fuels and solvents. In addition, with genetic engineering technology, bacteria can be programmed to make various substances used in food science, agriculture and medicine. The genetic systems of bacteria are the foundation of the biotechnology industry discussed below.

In the foods industry, lactic acid bacteria such as *Lactobacillus, Lactococcus* and *Streptococcus* are used in the manufacture of dairy products such as cheeses, including cottage cheese and cream cheese, cultured butter, sour cream, buttermilk, yogurt and kefir. Lactic acid bacteria and acetic acid bacteria are used in pickling processes such as olives, cucumber pickles and sauerkraut. Bacterial fermentations are used in processing of teas, coffee, cocoa, soy sauce, sausages and an amazing variety of foods in our everyday lives.

In the pharmaceutical industry, bacteria are used to produce antibiotics, vaccines, and medically-useful enzymes. Most antibiotics are made by bacteria that live in soil. Actinomycetes such as *Streptomyces* produce tetracyclines, erythromycin, streptomycin, rifamycin and ivermectin. *Bacillus* and *Paenibacillus* species produce bacitracin and polymyxin.

Bacterial products are used in the manufacture of vaccines for immunisation against infectious disease. Vaccines against diphtheria, whooping cough, tetanus, typhoid fever and cholera are made from components of the bacteria that cause the respective diseases.

It is significant to note here that the use of antibiotics against infectious disease and the widespread practice of vaccination (immunisation) against infectious disease are two twentieth-century developments that have drastically increased the quality of life and the average life expectancy of individuals in developed countries.

BIOTECHNOLOGY

The biotechnology industry uses bacterial cells for the production of biological substances that are useful to human existence, including fuels, foods, medicines, hormones, enzymes, proteins, and nucleic acids.

The possibilities of biotechnology are endless considering the gene reservoirs and genetic capabilities within the bacteria. Pasteur said it best, "Never underestimate the power of the microbe."

Biotechnology has produced human hormones such as insulin, enzymes such as streptokinase, and human proteins such as interferon and tumour necrosis factor. These products are used for the treatment of a various medical conditions and diseases including diabetes, heart attack, tuberculosis, AIDS and SLE. Botulinum toxin and BT insecticide are bacterial products used in medicine and pest control, respectively One biotechnological application of bacteria involves the genetic construction of super strains of organisms to perform particular metabolic tasks in the environment. For example, bacteria which have been engineered genetically to degrade petroleum products are used in cleanup of oil spills and other bioremediation efforts.

Another area of biotechnology involves improvement of the qualities of plants through genetic engineering. Genes can be introduced into plants by a bacterium *Agrobacterium tumefaciens*. Using *A. tumefaciens*, plants have been genetically engineered so that they are resistant to certain pests, herbicides, and diseases.

Finally, it should not be overlooked that industrial, pharmaceutical and food microbiology are applications of biotechnology. Archaea and bacteria are involved in production of biofuels.

Bacteria are the main producers of clinically useful antibiotics; they are a source of vaccines against once dreaded diseases; they are probiotics that enhance our health; and they are primary participants in the fermentations of dairy products and many other foods.

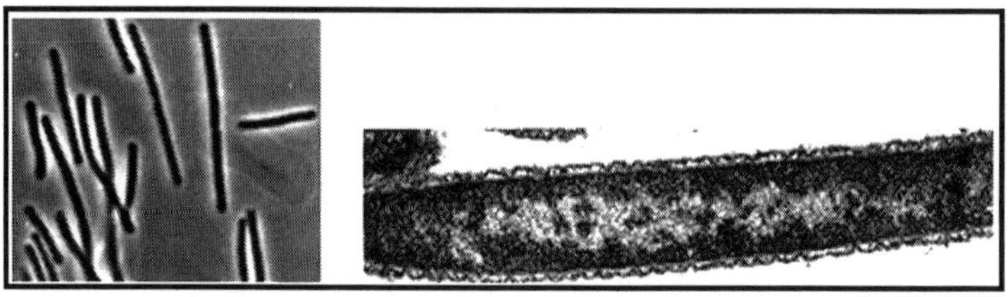

Fig. *Thermus Aquaticus, the* Thermophilic Bacterium that is the Source of Taq Polymerase. L wet Mount; R electron Micrograph. T.D. Brock. Life at High Temperatures. The Polymerase Chain Reaction (PCR), a Mainstay of the Biotechnology Industry because it Allows Duplication of Genes Starting with a Single Molecule of DNA, is Based on the Use of the DNA Polymerase Enzyme Derived from Thermus Aquaticus.

NUTRITIONAL REQUIREMENTS OF CELLS

Every organism must find in its environment all of the substances required for energy generation and cellular biosynthesis. The chemicals and elements of this environment that are utilised for bacterial growth are referred to as nutrients or nutritional requirements. Many bacteria can be grown the laboratory in culture media which are designed to provide all the essential nutrients in solution for bacterial growth.

Bacteria that are symbionts or obligate intracellular parasites of other cells, usually eucaryotic cells, are (not unexpectedly) difficult to grow outside of their natural host cells. Whether the microbe is a mutualist or parasite, the host cell must ultimately provide the nutritional requirements of its resident. Many bacteria can be identified in the environment by inspection or using genetic techniques, but attempts to isolate and grow them in artificial culture has been unsuccessful. This, in part, is the basis of the estimate that we may know less than one percent of all procaryotes that exist.

THE MAJOR ELEMENTS

At an elementary level, the nutritional requirements of a bacterium such as *E. coli* are revealed by the cell's elemental composition, which consists of C, H, O, N, S. P, K, Mg, Fe, Ca, Mn, and traces of Zn, Co, Cu, and Mo. These elements are found in the form of water, inorganic ions, small molecules, and macromolecules which serve either a structural or functional role in the cells.

Table. Major Elements, their Sources and Functions in Bacterial Cells.

Element	Per cent of Dry Weight	Source	Function
Carbon	50	Organic compounds or CO_2	Main constituent of cellular material
Oxygen	20	H_2O, organic compounds, CO_2, and O_2	Constituent of cell material and cell water; O_2 is electron acceptor in aerobic respiration
Nitrogen	14	NH_3, NO_3, organic compounds, N_2	Constituent of amino acids, nucleic acids nucleotides, and coenzymes
Hydrogen	8	H_2O, organic compounds, H_2	Main constituent of organic compounds and cell water
Phosphorus	3	inorganic phosphates (PO_4)	Constituent of nucleic acids, nucleotides, phospholipids, LPS, teichoic acids
Sulfur	1	SO_4, H_2S, S^o, organic sulfur compounds	Constituent of cysteine, methionine, glutathione, several coenzymes
Potassium	1	Potassium salts	Main cellular inorganic cation and cofactor for certain enzymes
Magnesium	0.5	Magnesium salts	Inorganic cellular cation, cofactor for certain enzymatic reactions
Calcium	0.5	Calcium salts	Inorganic cellular cation, cofactor for certain enzymes and a component of endospores
Iron	0.2	Iron salts	Component of cytochromes and certain nonheme iron-proteins and a cofactor for some enzymatic reactions

TRACE ELEMENTS

Trace elements are metal ions required by certain cells in such small amounts that it is difficult to detect (measure) them, and it is not necessary to add them to culture media as nutrients. Trace elements are required in such small amounts that they are present as "contaminants" of the water or other media components. As metal ions, the trace elements usually act as cofactors for essential enzymatic reactions in the cell. One organism's trace element may be another's required element and vice-versa, but the usual cations that qualify as trace elements in bacterial nutrition are Mn, Co, Zn, Cu, and Mo.

CARBON AND ENERGY SOURCES FOR BACTERIAL GROWTH

In order to grow in nature or in the laboratory, a bacterium must have an energy source, a source of carbon and other required nutrients, and a permissive range of physical conditions such as O_2 concentration, temperature, and pH. Sometimes bacteria are referred to as individuals or groups based on their patterns of growth under various chemical (nutritional) or physical conditions. For example, phototrophs are organisms that use light as an energy source; anaerobes are organisms that grow without oxygen; thermophiles are organisms that grow at high temperatures. All living organisms require a source of energy. Organisms that use radiant energy (light) are called phototrophs. Organisms that use (oxidise) an organic form of carbon are called heterotrophs or (chemo)heterotrophs. Organisms that oxidise inorganic compounds are called lithotrophs.

The carbon requirements of organisms must be met by organic carbon (a chemical compound with a carbon-hydrogen bond) or by CO_2. Organisms that use organic carbon are heterotrophs and organisms that use CO_2 as a sole source of carbon for growth are called autotrophs. Thus, on the basis of carbon and energy sources for growth four major nutritional types of procaryotes may be defined.

Table. Major Nutritional Types of Procaryotes.

Nutritional Type	Energy Source	Carbon Source	Examples
Photoautotrophs	Light	CO_2	Cyanobacteria, some Purple and Green Bacteria
Photoheterotrophs	Light	Organic compounds	Some Purple and Green Bacteria
Chemoautotrophs or Lithotrophs (Lithoautotrophs)	Inorganic compounds, e.g. H_2, NH_3, NO_2, H_2S	CO_2	A few Bacteria and many Archaea
Chemoheterotrophs or Heterotrophs	Organic compounds	Organic compounds	Most Bacteria, some Archaea

Almost all eucaryotes are either photoautotrophic (*e.g.*, plants and algae) or heterotrophic (*e.g.*, animals, protozoa, fungi). Lithotrophy is unique to procaryotes and photoheterotrophy, common in the Purple and Green Bacteria, occurs only in a very few eucaryotic algae. Phototrophy has not been found in the Archaea, except for non-photosynthetic light-driven ATP synthesis in the extreme halophiles.

GROWTH FACTORS

This simplified scheme for use of carbon, either organic carbon or CO_2, ignores the possibility that an organism, whether it is an autotroph or a heterotroph, may require small amounts of certain organic compounds for growth because they are essential substances that the organism is unable to synthesize from available nutrients. Such compounds are called growth factors. Growth factors are required in small amounts by cells because they fulfil specific roles in biosynthesis. The need for a growth factor results from either a blocked or missing metabolic pathway in the cells.

Growth factors are organised into three categories.

1. *Purines and Pyrimidines:* Required for synthesis of nucleic acids (DNA and RNA)
2. *Amino Acids*: Required for the synthesis of proteins
3. *Vitamins:* Needed as coenzymes and functional groups of certain enzymes

Table. Common Vitamins Required in the Nutrition of Certain Bacteria.

Vitamin	Coenzyme Form	Function
p-Aminobenzoic acid (PABA)	-	Precursor for the biosynthesis of folic acid
Folic acid	Tetrahydrofolate	Transfer of one-carbon units and required for synthesis of thymine, purine bases, serine, methionine and pantothenate
Biotin	Biotin	Biosynthetic reactions that require CO_2 fixation
Lipoic acid	Lipoamide	Transfer of acyl groups in oxidation of keto acids
Mercaptoethane-sulfonic acid	Coenzyme M	CH_4 production by methanogens
Nicotinic acid	NAD (nicotinamide adenine dinucleotide) and NADP	Electron carrier in dehydrogenation reactions
Pantothenic acid	Coenzyme A and the Acyl Carrier Protein (ACP)	Oxidation of keto acids and acyl group carriers in metabolism
Pyridoxine (B_6)	Pyridoxal phosphate	Transamination, deamination, decarboxylation and racemation of amino acids
Riboflavin (B_2)	FMN (flavin mononucleotide) and FAD (flavin adenine dinucleotide)	Oxidoreduction reactions
Thiamine (B_1)	Thiamine pyrophosphate (TPP)	Decarboxylation of keto acids and transaminase reactions
Vitamin B_{12}	Cobalamine coupled to adenine nucleoside	Transfer of methyl groups
Vitamin K	Quinones and napthoquinones	Electron transport processes

Some bacteria (*e.g. E. coli*) do not require any growth factors: they can synthesize all essential purines, pyrimidines, amino acids and vitamins, starting with their carbon source, as part of their own intermediary metabolism. Certain other bacteria (*e.g. Lactobacillus*) require purines, pyrimidines, vitamins and several amino acids in order to grow. These compounds must be added in advance to culture media that are used to grow these bacteria. The growth factors are not metabolised directly as sources of carbon or energy, rather they are assimilated by cells to fulfil their specific role in metabolism. Mutant strains of bacteria that require some growth factor not needed by the wild type (parent) strain are referred to as auxotrophs. Thus, a strain of *E. coli* that requires the amino acid tryptophan in order to grow would be called a tryptophan auxotroph and would be designated *E. coli trp*$^-$

Fig. Cross-feeding between *Staphylococcus Aureus* and *Haemophilus Influenzae* Growing on Blood Agar.

The function(s) of these vitamins in essential enzymatic reactions gives a clue why, if the cell cannot make the vitamin, it must be provided exogenously in order for growth to occur.

CULTURE MEDIA FOR THE GROWTH OF BACTERIA

For any bacterium to be propagated for any purpose it is necessary to provide the appropriate biochemical and biophysical environment. The biochemical (nutritional) environment is made available as a culture medium, and depending upon the special needs of particular bacteria (as well as particular investigators) a large variety and types of culture media have been developed with different purposes and uses. Culture media are employed in the isolation and maintenance of pure cultures of bacteria and are also used for identification of bacteria according to their biochemical and physiological properties.

The manner in which bacteria are cultivated, and the purpose of culture media, varies widely. Liquid media are used for growth of pure batch cultures, while solidified media are used widely for the isolation of pure cultures, for estimating viable bacterial populations, and a variety of other purposes. The usual gelling agent for solid or semisolid medium is agar, a hydrocolloid derived from red algae. Agar is used because of its unique physical properties (it melts at 100°C and remains liquid until cooled to 40°C, the temperature at which it gels) and because it cannot be metabolised by most bacteria. Hence as a medium component it is relatively inert; it simply holds (gels) nutrients that are in aqueous solution.

Types of Culture Media

Culture media may be classified into several categories depending on their composition or use. A chemically-defined (synthetic) medium is one in which the exact chemical composition is known. A complex (undefined) medium is one in which the exact chemical constitution of the medium is not known. Defined media are usually composed of pure biochemicals off the shelf; complex media usually contain complex materials of biological origin such as blood or milk or yeast extract or beef extract, the exact chemical composition of which is obviously undetermined.

A defined medium is a minimal medium if it provides only the exact nutrients (including any growth factors) needed by the organism for growth. The use of defined minimal media requires the investigator to know the exact nutritional requirements of the organisms in question. Chemically-defined media are of value in studying the minimal nutritional requirements of microorganisms, for enrichment cultures, and for a wide variety of physiological studies. Complex media usually provide the full range of growth factors that may be required by an organism so they may be more handily used to cultivate unknown bacteria or bacteria whose nutritional requirement are complex (*i.e.*, organisms that require a lot of growth factors, known or unknown). Complex media are usually used for cultivation of bacterial pathogens and other fastidious bacteria.

Most pathogenic bacteria of animals, which have adapted themselves to growth in animal tissues, require complex media for their growth. Blood, serum and tissue extracts are frequently added to culture media for the cultivation of pathogens. Even so, for a few fastidious pathogens such as *Treponema pallidum*, the agent of syphilis, and *Mycobacterium leprae*, the cause of leprosy, artificial culture media and conditions have not been established. This fact thwarts the ability to do basic research on these pathogens and the diseases that they cause.

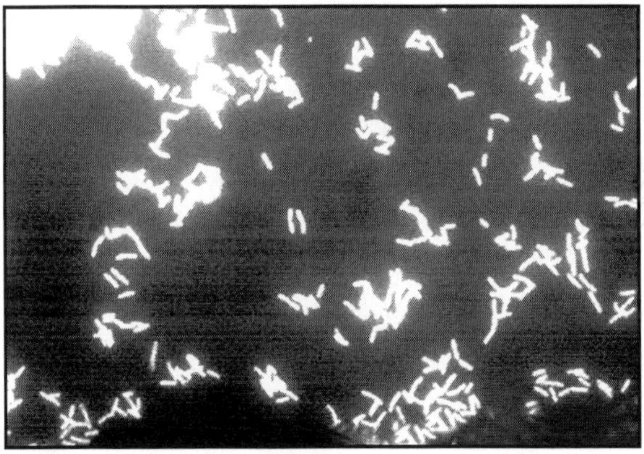

Fig. *Legionella Pneumophila.* Direct Fluorescent antibody (DFA) Stain of a Patient Respiratory Tract Specimen.

Other concepts employed in the construction of culture media are the principles of selection and enrichment. A selective medium is one which has a component(s) added to it which will inhibit or prevent the growth of certain types or species of bacteria and/or promote the growth of desired species. One can also adjust the physical conditions of a culture medium, such as pH and temperature, to render it selective for organisms that are able to grow under these certain conditions.

A culture medium may also be a differential medium if allows the investigator to distinguish between different types of bacteria based on some observable trait in their pattern of growth on the medium. Thus a selective, differential medium for the isolation of *Staphylococcus aureus,* the most common bacterial pathogen of humans, contains a very high concentration of salt (which the staph will tolerate) that inhibits most other bacteria, mannitol as a source of fermentable sugar, and a pH indicator dye. From clinical specimens, only staph will grow. *S. aureus* is differentiated from *S. epidermidis* (a non-pathogenic component of the normal flora) on the basis of its ability to ferment mannitol. Mannitol-fermenting colonies *(S. aureus) produce* acid which reacts with the indicator dye forming a coloured halo around the colonies; mannitol non-fermenters (*S. epidermidis*) use other non-fermentative substrates in the medium for growth and do not form a halo around their colonies.

An enrichment medium employs a slightly different twist. An enrichment medium contains some component that permits the growth of specific types or species of bacteria, usually because they alone can utilise the component from their environment. However, an enrichment medium may have selective features. An enrichment medium for non-symbiotic nitrogen-fixing bacteria omits a source of added nitrogen to the medium. The medium is inoculated with a potential source of these bacteria (*e.g.*, a soil sample) and incubated in the atmosphere wherein the only source of nitrogen available is N_2. A selective enrichment medium for growth of the extreme halophile (*Halococcus*) contains nearly 25 percent salt [NaCl], which is required by the extreme halophile and which inhibits the growth of all other procaryotes.

Table. Minimal Medium for the Growth of *Bacillus Megaterium*. An Example of a Chemically-defined Medium for Growth of a Heterotrophic Bacterium.

Component	Amount	Function of Component
Sucrose	10.0 g	C and energy source
K_2HPO_4	2.5 g	pH buffer; P and K source
KH_2PO_4	2.5 g	pH buffer; P and K source
$(NH_4)2HPO_4$	1.0 g	pH buffer; N and P source
$MgSO_4$ $7H_2O$	0.20 g	S and Mg^{++} source
$FeSO_4$ $7H_2O$	0.01 g	Fe^{++} source
$MnSO_4$ $7H_2O$	0.007 g	Mn^{++} Source
Water	985 ml	
pH 7.0		

Table. Defined Medium (Also an Enrichment Medium) for the Growth of *Thiobacillus Thiooxidans*, a Lithoautotrophic Bacterium.

Component	Amount	Function of Component
NH_4Cl	0.52 g	N source
KH_2PO_4	0.28 g	P and K source
$MgSO_4$ $7H_2O$	0.25 g	S and Mg^{++} source
$CaCl_2$ $2H_2O$	0.07 g	Ca^{++} source
Elemental Sulfur	1.56 g	Energy source
CO_2	5%*	C source
Water	1000 ml	
pH 3.0		

Note:

* Aerate medium intermittently with air containing 5 per cent CO2.

Table. Complex Medium for the Growth of Fastidious Bacteria.

Component	Amount	Function of Component
Beef extract	1.5 g	Source of vitamins and other growth factors
Yeast extract	3.0 g	Source of vitamins and other growth factors
Peptone	6.0 g	Source of amino acids, N, S, and P
Glucose	1.0 g	C and energy source
Agar	15.0 g	Inert solidifying agent
Water	1000 ml	
pH 6.6		

Table. Selective Enrichment Medium for Growth of Extreme Halophiles.

Component	Amount	Function of Component
Casamino acids	7.5 g	Source of amino acids, N, S and P
Yeast extract	10.0 g	Source of growth factors
Trisodium citrate	3.0 g	C and energy source
KCl	2.0 g	K^+ source
$MgSO_4$ 7 H_2O	20.0 g	S and Mg^{++} source
$FeCl_2$	0.023 g	Fe^{++} source
NaCl	250 g	Na^+ source for halophiles and inhibitory to nonhalophiles
Water	1000 ml	
pH 7.4		

PHYSICAL AND ENVIRONMENTAL REQUIREMENTS FOR MICROBIAL GROWTH

The procaryotes exist in nature under an enormous range of physical conditions such as O_2 concentration, Hydrogen ion concentration (pH) and temperature. The exclusion limits of life on the planet, with regard to environmental parameters, are always set by some microorganism, most often a procaryote, and frequently an Archaeon. Applied to all microorganisms is a vocabulary of terms used to describe their growth (ability to grow) within a range of physical conditions. A thermophile grows at high temperatures, an acidiphile grows at low pH, an osmophile grows at high solute concentration, and so on. This nomenclature will be employed in this section to describe the response of the procaryotes to a variety of physical conditions.

THE EFFECT OF OXYGEN

Oxygen is a universal component of cells and is always provided in large amounts by H_2O. However, procaryotes display a wide range of responses to molecular oxygen O_2.

- Obligate aerobes require O_2 for growth; they use O_2 as a final electron acceptor in aerobic respiration.

- Obligate anaerobes (occasionally called aerophobes) do not need or use O_2 as a nutrient. In fact, O_2 is a toxic substance, which either kills or inhibits their growth. Obligate anaerobic procaryotes may live by fermentation, anaerobic respiration, bacterial photosynthesis, or the novel process of methanogenesis.
- Facultative anaerobes (or facultative aerobes) are organisms that can switch between aerobic and anaerobic types of metabolism. Under anaerobic conditions (no O_2) they grow by fermentation or anaerobic respiration, but in the presence of O_2 they switch to aerobic respiration.
- Aerotolerant anaerobes are bacteria with an exclusively anaerobic (fermentative) type of metabolism but they are insensitive to the presence of O_2. They live by fermentation alone whether or not O_2 is present in their environment.

Table. Terms Used to Describe O_2 Relations of Microorganisms.

		Environment	
Group	**Aerobic**	**Anaerobic**	**O_2 Effect**
Obligate Aerobe	Growth	No growth	Required (utilized for aerobic respiration)
Microaerophile	Growth if level not too high	No growth	Required but at levels below 0.2 atm
Obligate Anaerobe	No growth	Growth Toxic	
Facultative Anaerobe (Facultative Aerobe)	Growth	Growth	Not required for growth but utilized when available
Aerotolerant Anaerobe	Growth	Growth	Not required and not utilized

The response of an organism to O_2 in its environment depends upon the occurrence and distribution of various enzymes which react with O_2 and various oxygen radicals that are invariably generated by cells in the presence of O_2. All cells contain enzymes capable of reacting with O_2. For example, oxidations of flavoproteins by O_2 invariably result in the formation of H_2O_2 (peroxide) as one major product and small quantities of an even more toxic free radical, superoxide or $O_2^{\cdot-}$. Also, chlorophyll and other pigments in cells can react with O_2 in the presence of light and generate singlet oxygen, another radical form of oxygen which is a potent oxidising agent in biological systems.

In aerobes and aerotolerant anaerobes the potential for lethal accumulation of superoxide is prevented by the enzyme superoxide dismutase. All organisms which can live in the presence of O_2 (whether or not they utilise it in their metabolism) contain superoxide dismutase. Nearly all organisms contain the enzyme catalase, which decomposes H_2O_2. Even though certain aerotolerant bacteria such as the lactic acid bacteria lack catalase, they decompose H_2O_2 by means of peroxidase enzymes which derive electrons from $NADH_2$ to reduce peroxide to H_2O. Obligate anaerobes lack superoxide dismutase and catalase and/or peroxidase, and therefore undergo lethal oxidations by various oxygen radicals when they are exposed to O_2.

All photosynthetic (and some non-photosynthetic) organisms are protected from lethal oxidations of singlet oxygen by their possession of carotenoid pigments which physically react with the singlet oxygen radical and lower it to its non-toxic "ground" (triplet) state. Carotenoids are said to "quench" singlet oxygen radicals.

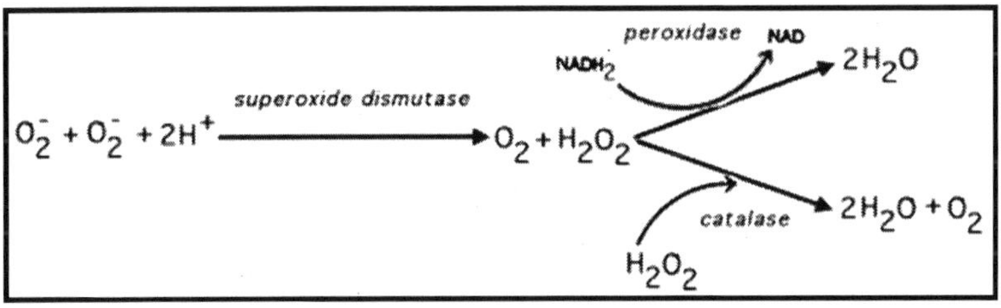

Fig. The Action of Superoxide Dismutase, Catalase and Peroxidase. These Enzymes Detoxify Oxygen Radicals that are Inevitably Generated by Living Systems in the Presence of O_2. The Distribution of these Enzymes in Cells Determines their Ability to Exist in the Presence of O_2

Table. Distribution of Superoxide Dismutase, Catalase and Peroxidase in Procaryotes with Different O_2 Tolerances.

Group	Superoxide Dismutase	Catalase	Peroxidase
Obligate aerobes and most facultative anaerobes (e.g. Enterics)	+	+	–
Most aerotolerant anaerobes (e.g. Streptococci)	+	–	+
Obligate anaerobes (e.g. Clostridia, Methanogens, Bacteroides)	–	–	–

THE EFFECT OF PH ON GROWTH

The pH, or hydrogen ion concentration, $[H^+]$, of natural environments varies from about 0.5 in the most acidic soils to about 10.5 in the most alkaline lakes. Appreciating that pH is measured on a logarithmic scale, the $[H^+]$ of natural environments varies over a billion-fold and some microorganisms are living at the extremes, as well as every point between the extremes! Most free-living procaryotes can grow over a range of 3 pH units, about a thousand fold change in $[H^+]$.

The range of pH over which an organism grows is defined by three cardinal points: the minimum pH, below which the organism cannot grow, the maximum pH, above which the organism cannot grow, and the optimum pH, at which the organism grows best. For most bacteria there is an orderly increase in growth rate between the minimum and the optimum and a corresponding orderly decrease in growth rate between the optimum and the maximum pH, reflecting the general effect of changing [H^+] on the rates of enzymatic reaction.

Microorganisms which grow at an optimum pH well below neutrality (7.0) are called acidophiles. Those which grow best at neutral pH are called neutrophiles and those that grow best under alkaline conditions are called alkaliphiles. Obligate acidophiles, such as some *Thiobacillus* species, actually require a low pH for growth since their membranes dissolve and the cells lyse at neutrality. Several genera of Archaea, including *Sulfolobus* and *Thermoplasma*, are obligate acidophiles. Among eukaryotes, many fungi are acidophiles, but the champion of growth at low pH is the eucaryotic alga *Cyanidium* which can grow at a pH of 0.

In the construction and use of culture media, one must always consider the optimum pH for growth of a desired organism and incorporate buffers in order to maintain the pH of the medium in the changing milieu of bacterial waste products that accumulate during growth. Many pathogenic bacteria exhibit a relatively narrow range of pH over which they will grow. Most diagnostic media for the growth and identification of human pathogens have a pH near 7.

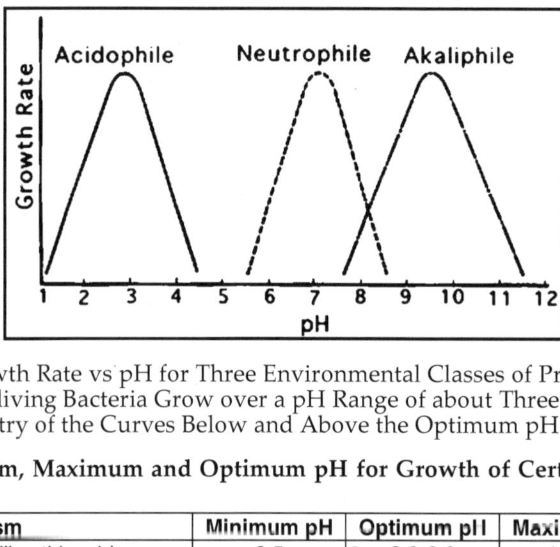

Fig. Growth Rate vs pH for Three Environmental Classes of Procaryotes. Most Free-living Bacteria Grow over a pH Range of about Three Units. Note the Symmetry of the Curves Below and Above the Optimum pH for Growth.

Table. Minimum, Maximum and Optimum pH for Growth of Certain Procaryotes.

Organism	Minimum pH	Optimum pH	Maximum pH
Thiobacillus thiooxidans	0.5	2.0-2.8	4.0-6.0
Sulfolobus acidocaldarius	1.0	2.0-3.0	5.0
Bacillus acidocaldarius	2.0	4.0	6.0
Zymomonas lindneri	3.5	5.5-6.0	7.5
Lactobacillus acidophilus	4.0-4.6	5.8-6.6	6.8
Staphylococcus aureus	4.2	7.0-7.5	9.3
Escherichia coli	4.4	6.0-7.0	9.0
Clostridium sporogenes	5.0-5.8	6.0-7.6	8.5-9.0
Erwinia caratovora	5.6	7.1	9.3
Pseudomonas aeruginosa	5.6	6.6-7.0	8.0
Thiobacillus novellus	5.7	7.0	9.0
Streptococcus pneumoniae	6.5	7.8	8.3
Nitrobacter sp	6.6	7.6-8.6	10.0

THE EFFECT OF TEMPERATURE ON GROWTH

Microorganisms have been found growing in virtually all environments where there is liquid water, regardless of its temperature. In 1966, Professor Thomas D. Brock, then at Indiana University, made the amazing discovery in boiling hot springs of Yellowstone National Park that bacteria were not just surviving there, they were growing and flourishing. Brock's discovery of thermophilic bacteria, archaea and other "extremophiles" in Yellowstone is summarised for the general public in an article at this web site.

Subsequently, procaryotes have been detected growing around black smokers and hydrothermal vents in the deep sea at temperatures at least as high as 120 degrees. Microorganisms have been found growing at very low temperatures as well. In supercooled solutions of H_2O as low as -20 degrees, certain organisms can extract water for growth, and many forms of life flourish in the icy waters of the Antarctic, as well as household refrigerators, near 0 degrees.

A particular microorganism will exhibit a range of temperature over which it can grow, defined by three cardinal points in the same manner as pH. Considering the total span of temperature where liquid water exists, the procaryotes may be subdivided into several subclasses on the basis of one or another of their cardinal points for growth. For example, organisms with an optimum temperature near 37 degrees (the body temperature of warm-blooded animals) are called mesophiles. Organisms with an optimum T between about 45 degrees and 70 degrees are thermophiles. Some Archaea with an optimum T of 80 degrees or higher and a maximum T as high as 115 degrees, are now referred to as extreme thermophiles or hyperthermophiles. The cold-loving organisms are psychrophiles defined by their ability to grow at 0 degrees. A variant of a psychrophile (which usually has an optimum T of 10-15 degrees) is a psychrotroph, which grows at 0 degrees but displays an optimum T in the mesophile range, nearer room temperature.

Psychrotrophs are the scourge of food storage in refrigerators since they are invariably brought in from their mesophilic habitats and continue to grow in the refrigerated environment where they spoil the food. Of course, they grow slower at 2 degrees than at 25 degrees. Think how fast milk spoils on the counter top versus in the refrigerator.

Psychrophilic bacteria are adapted to their cool environment by having largely unsaturated fatty acids in their plasma membranes. Some psychrophiles, particularly those from the Antarctic have been found to contain polyunsaturated fatty acids, which generally do not occur in procaryotes. The degree of unsaturation of a fatty acid correlates with its solidification T or thermal transition stage (*i.e.*, the temperature at which the lipid melts or solidifies); unsaturated fatty acids remain liquid at low T but are also denatured at moderate T; saturated fatty acids, as in the membranes of thermophilic bacteria, are stable at high temperatures, but they also solidify at relatively high T. Thus, saturated fatty acids (like butter) are solid at room temperature while unsaturated fatty acids (like safflower oil) remain liquid in the refrigerator. Whether fatty acids in a membrane are in a liquid or a solid phase affects the fluidity of the membrane, which directly affects its ability to function. Psychrophiles also have enzymes that continue to function, albeit at a reduced rate, at temperatures at or near 0 degrees.

Usually, psychrophile proteins and/or membranes, which adapt them to low temperatures, do not function at the body temperatures of warm-blooded animals (37 degrees) so that they are unable to grow at even moderate temperatures.

Thermophiles are adapted to temperatures above 60 degrees in a variety of ways. Often thermophiles have a high G + C content in their DNA such that the melting point of the DNA (the temperature at which the strands of the double helix separate) is at least as high as the organism's maximum T for growth. But this is not always the case, and the correlation is far from perfect, so thermophile DNA must be stabilised in these cells by other means.

The membrane fatty acids of thermophilic bacteria are highly saturated allowing their membranes to remain stable and functional at high temperatures. The membranes of hyperthermophiles, virtually all of which are Archaea, are not composed of fatty acids but of repeating subunits of the C5 compound, phytane, a branched, saturated, "isoprenoid" substance, which contributes heavily to the ability of these bacteria to live in superheated environments. The structural proteins (*e.g.*, ribosomal proteins, transport proteins (permeases) and enzymes of thermophiles and hyperthermophiles are very heat stable compared with their mesophilic counterparts. The proteins are modified in a number of ways including dehydration and through slight changes in their primary structure, which accounts for their thermal stability.

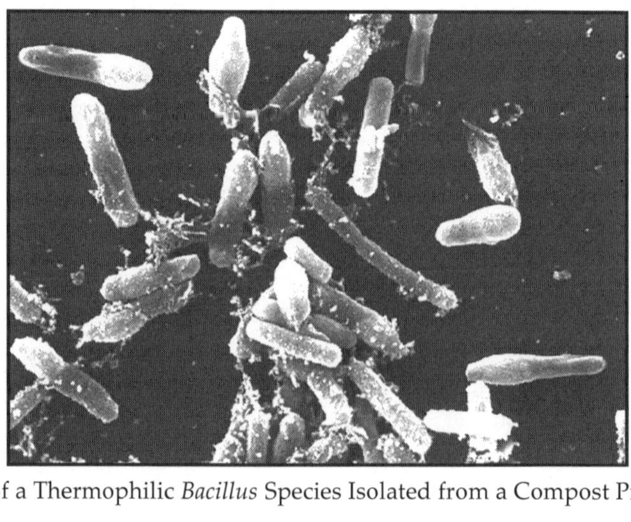

Fig. SEM of a Thermophilic *Bacillus* Species Isolated from a Compost Pile at 55° C.

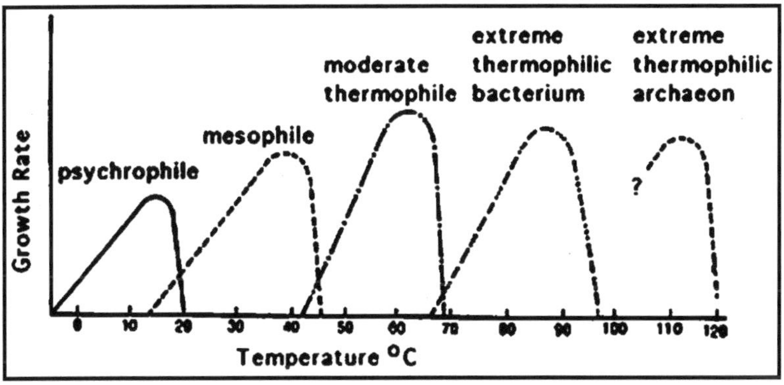

Fig. Growth Rate vs Temperature for Five Environmental Classes of Procaryotes. Most Procaryotes will Grow over a Temperature Range of about 30 Degrees. The Curves Exhibit Three Cardinal Points: Minimum, Optimum and Maximum Temperatures for Growth. There is a Steady Increase in Growth Rate between the Minimum and Optimum Temperatures, but Slightly Past the Optimum a Critical Thermolabile Cellular Event Occurs, and the Growth Rates Plunge Rapidly as the Maximum T is Approached.

Table. Terms Used to Describe Microorganisms in Relation to Temperature Requirements for Growth.

Group	Minimum	Optimum	Maximum	Comments
Psychrophile	Below 0	10-15	Below 20	Grow best at relatively low T
Psychrotroph	0	15-30	Above 25	Able to grow at low T but prefer moderate T
Mesophile	10-15	30-40	Below 45	Most bacteria esp. those living in association with warm-blooded animals
Thermophile*	45	50-85	Above 100 (boiling)	Among all thermophiles is wide variation in optimum and maximum T

Table. Minimum, Maximum and Optimum Temperature for Growth of Certain Bacteria and Archaea.

Bacterium	Minimum	Optimum	Maximum
	Temperature for growth (degrees C)		
Listeria monocytogenes	1	30-37	45
Vibrio marinus	4	15	30
Pseudomonas maltophilia	4	35	41
Thiobacillus novellus	5	25-30	42
Staphylococcus aureus	10	30-37	45
Escherichia coli	10	37	45
Clostridium kluyveri	19	35	37
Streptococcus pyogenes	20	37	40
Streptococcus pneumoniae	25	37	42
Bacillus flavothermus	30	60	72
Thermus aquaticus	40	70-72	79
Methanococcus jannaschii	60	85	90
Sulfolobus acidocaldarius	70	75-85	90
Pyrobacterium brockii	80	102-105	115

Table. Optimum Growth Temperature of Some Procaryotes.

Genus and Species	Optimal Growth Temp (Degrees C)
Vibrio cholerae	18-37
Photobacterium phosphoreum	20
Rhizobium leguminosarum	20
Streptomyces griseus	25
Rhodobacter sphaeroides	25-30
Pseudomonas fluorescens	25-30
Erwinia amylovora	27-30
Staphylococcus aureus	30-37
Escherichia coli	37
Mycobacterium tuberculosis	37
Pseudomonas aeruginosa	37
Streptococcus pyogenes	37
Treponema pallidum	37
Thermoplasma acidophilum	59
Thermus aquaticus	70
Bacillus caldolyticus	72
Pyrococcus furiosus	100

Table. Hyperthermophilic Archaea.

	Temperature for Growth (Degrees C)			
Genus pH	Minimum	Optimum	Maximum	Optimum
Sulfolobus	55	75-85	87	2-3
Desulfurococcus	60	85	93	6
Methanothermus	60	83	88	6-7
Pyrodictium	82	105	113	6
Methanopyrus	85	100	110	7

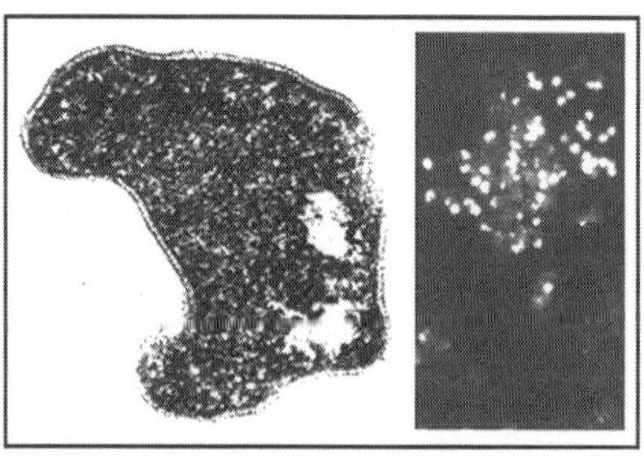

Fig. *Sulfolobus Acidocaldarius* is An Extreme Thermophile and An Acidophile Found in Geothermally-heated Acid Springs, Mud Pots and Surface Soils with Temperatures from 60 to 95 Degrees C, and a pH of 1 to 5. Left: Electron Micrograph of a Thin Section (85,000X). Under the Electron Microscope the Organism Appears as Irregular Spheres which are Often Lobed. Right: Fluorescent Photomicrograph of Cells Attached to a Sulphur Crystal. Fimbrial-like Appendages have been Observed on the Cells Attached to Solid Surfaces such as Sulphur Crystals.

Water is the solvent in which the molecules of life are dissolved, and the availability of water is therefore a critical factor that affects the growth of all cells. The availability of water for a cell depends upon its presence in the atmosphere (relative humidity) or its presence in solution or a substance (water activity). The water activity (A_w) of pure H_2O is 1.0 (100 per cent water). Water activity is affected by the presence of solutes such as salts or sugars, that are dissolved in the water. The higher the solute concentration of a substance, the lower is the water activity and vice-versa. Microorganisms live over a range of A_w from 1.0 to 0.7. The A_w of human blood is 0.99; seawater = 0.98; maple syrup = 0.90; Great Salt Lake = 0.75. Water activities in agricultural soils range between 0.9 and 1.0.

The only common solute in nature that occurs over a wide concentration range is salt [NaCl], and some microorganisms are named based on their growth response to salt. Microorganisms that require some NaCl for growth are halophiles. Mild halophiles require 1-6 per cent salt, moderate halophiles require 6-15 per cent salt; extreme halophiles that require 15-30 per cent NaCl for growth are found among the archaea. Bacteria that are able to grow at moderate salt concentrations, even though they grow best in the absence of NaCl, are called halotolerant. Although halophiles are "osmophiles" (and halotolerant organisms are "osmotolerant") the term osmophiles is usually reserved for organisms that are able to live in environments high in sugar. Organisms which live in dry environments (made dry by lack of water) are called xerophiles.

The concept of lowering water activity in order to prevent bacterial growth is the basis for preservation of foods by drying (in sunlight or by evaporation) or by addition of high concentrations of salt or sugar.

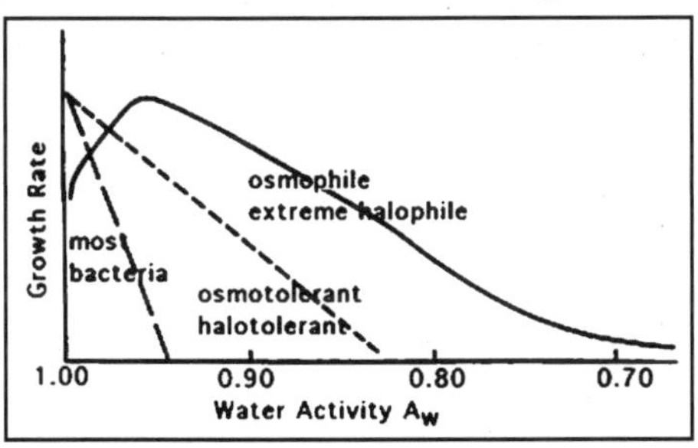

Fig. Growth Rate vs Osmolarity for Different Classes of Procaryotes. Osmolarity is Determined by Solute Concentration in the Environment. Osmolarity is Inversely Related to Water Activity (A_w), which is More Like a Measure of the Concentration of Water (H_2O) in a Solution. Increased Solute Concentration Means Increased Osmolarity and Decreased A_w. From Left to Right the Graph Shows the Growth Rate of a Normal (Non-halophile) such as *E. Coli* or *Pseudomonas,* the Growth Rate of a Halotolerant Bacterium such as *Staphylococcus Aureus, and* the Growth Rate of an Extreme Halophile such as the Archaean *Halococcus. Note* that a True Halophile Grows Best at Salt Concentrations where Most Bacteria are Inhibited.

Table. Limiting Water Activities (A_w) for Growth of Certain Procaryotes.

Organism	Minimum A_w for Growth
Caulobacter	1.00
Spirillum	1.00
Pseudomonas	.91

Salmonella/E. coli	.91
Lactobacillus	.90
Bacillus	.90
Staphylococcus	.85
Halococcus	.75

MEASUREMENT OF BACTERIAL GROWTH

Growth is an orderly increase in the quantity of cellular constituents. It depends upon the ability of the cell to form new protoplasm from nutrients available in the environment.

In most bacteria, growth involves increase in cell mass and number of ribosomes, duplication of the bacterial chromosome, synthesis of new cell wall and plasma membrane, partitioning of the two chromosomes, septum formation, and cell division. This asexual process of reproduction is called binary fission.

For unicellular organisms such as the bacteria, growth can be measured in terms of two different parameters:

1. Changes in cell mass and
2. Changes in cell numbers.

METHODS FOR MEASUREMENT OF CELL MASS

Methods for measurement of the cell mass involve both direct and indirect techniques:

- Direct physical measurement of dry weight, wet weight, or volume of cells after centrifugation.
- Direct chemical measurement of some chemical component of the cells such as total N, total protein, or total DNA content.
- Indirect measurement of chemical activity such as rate of O_2 production or consumption, CO_2 production or consumption, etc.
- Turbidity measurements employ a variety of instruments to determine the amount of light scattered by a suspension of cells. Particulate objects such as bacteria scatter light in proportion to their numbers. The turbidity or optical density of a suspension of cells is directly related to cell mass or cell number, after construction and calibration of a standard curve.
- The method is simple and non-destructive, but the sensitivity is limited to about 10^7 cells per ml for most bacteria.

METHODS FOR MEASUREMENT OF CELL NUMBERS

Measuring techniques involve direct counts, visually or instrumentally, and indirect viable cell counts:

- Direct microscopic counts are possible using special slides known as counting chambers. Dead cells cannot be distinguished from living ones. Only dense suspensions can be counted (>107 cells per ml), but samples can be concentrated by centrifugation or filtration to increase sensitivity.

A variation of the direct microscopic count has been used to observe and measure growth of bacteria in natural environments. In order to detect and prove that thermophilic bacteria were growing in boiling hot springs, T.D. Brock immersed microscope slides in the springs and withdrew them periodically for microscopic observation. The bacteria in the boiling water attached to the glass slides naturally and grew as microcolonies on the surface.

- Electronic counting chambers count numbers and measure size distribution of cells. For cells the size of bacteria the suspending medium must be very clean. Such electronic devices are more often used to count eucaryotic cells such as blood cells.
- Indirect viable cell counts, also called plate counts, involve plating out (spreading) a sample of a culture on a nutrient agar surface. The sample or cell suspension can be diluted in a non-toxic diluent (*e.g.*, water or saline) before plating. If plated on a suitable medium, each viable unit grows and forms a colony. Each colony that can be counted is called a colony forming unit (cfu) and the number of cfu's is related to the viable number of bacteria in the sample.

Table. Some Methods Used to Measure Bacterial Growth.

Method	Application	Comments
Direct microscopic count	Enumeration of bacteria in milk or cellular vaccines	Cannot distinguish living from nonliving cells
Viable cell count (colony counts)	Enumeration of bacteria in milk, foods, soil, water, laboratory cultures, etc.	Very sensitive if plating conditions are optimal
Turbidity measurement	Estimations of large numbers of bacteria in clear liquid media and broths	Fast and nondestructive, but cannot detect cell densities less than 10^7 cells per ml
Measurement of total N or protein	Measurement of total cell yield from very dense cultures	only practical application is in the research laboratory
Measurement of Biochemical activity e.g. O2 uptake CO2 production, ATP production, etc.	Microbiological assays	Requires a fixed standard to relate chemical activity to cell mass and/or cell numbers
Measurement of dry weight or wet weight of cells or volume of cells after centrifugation	Measurement of total cell yield in cultures	Probably more sensitive than total N or total protein measurements

Advantages of the technique are its sensitivity (theoretically, a single cell can be detected), and it allows for inspection and positive identification of the organism counted.

Disadvantages are:

- Only living cells develop colonies that are counted;
- Clumps or chains of cells develop into a single colony
- Colonies develop only from those organisms for which the cultural conditions are suitable for growth.

The latter makes the technique virtually useless to characterise or count the total number of bacteria in complex microbial ecosystems such as soil or the animal rumen or gastrointestinal tract. Genetic probes can be used to demonstrate the diversity and relative abundance of procaryotes in such an environment, but many species identified by genetic techniques have so far proven unculturable.

THE BACTERIAL GROWTH CURVE

In the laboratory, under favourable conditions, a growing bacterial population doubles at regular intervals. Growth is by geometric progression: 1, 2, 4, 8, etc., or $2^0, 2^1, 2^2, 2^3 ....2^n$ (where n = the number of generations). This is called exponential growth. In reality, exponential growth is only part of the bacterial life cycle, and not representative of the normal pattern of growth of bacteria in Nature. When a fresh medium is inoculated with a given number of cells, and the population growth is monitored over a period of time, plotting the data will yield a typical bacterial growth curve.

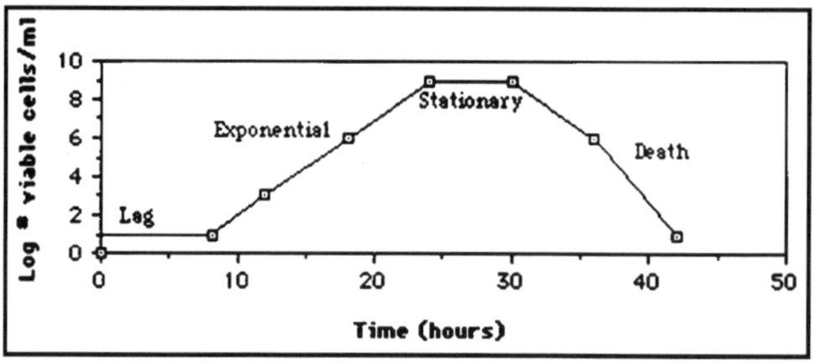

Fig. The Typical Bacterial Growth Curve. When Bacteria are Grown in a Closed System (Also called a Batch Culture), like a Test Tube, the Population of Cells Almost Always Exhibits these Growth Dynamics: Cells Initially Adjust to the New Medium (Lag Phase) until they can Start Dividing Regularly by the Process of Binary Fission (Exponential Phase). When their Growth becomes Limited, the Cells Stop Dividing (Stationary Phase), Until Eventually they Show Loss of Viability (Death Phase). Note the Parameters of the x and y Axes. Growth is Expressed as Change in the Number Viable Cells vs Time. Generation Times are Calculated During the Exponential Phase of Growth. Time Measurements are in Hours for Bacteria with Short Generation Times.

Four characteristic phases of the growth cycle are recognised:

1. *Lag Phase*: Immediately after inoculation of the cells into fresh medium, the population remains temporarily unchanged. Although there is no apparent cell division occurring, the cells may be growing in volume or mass, synthesizing enzymes, proteins, RNA, etc., and increasing in metabolic activity. The length of the lag phase is apparently dependent

on a wide variety of factors including the size of the inoculum; time necessary to recover from physical damage or shock in the transfer; time required for synthesis of essential coenzymes or division factors; and time required for synthesis of new (inducible) enzymes that are necessary to metabolise the substrates present in the medium.

2. *Exponential (Log) Phase:* The exponential phase of growth is a pattern of balanced growth wherein all the cells are dividing regularly by binary fission, and are growing by geometric progression. The cells divide at a constant rate depending upon the composition of the growth medium and the conditions of incubation. The rate of exponential growth of a bacterial culture is expressed as generation time, also the doubling time of the bacterial population. Generation time (G) is defined as the time (t) per generation (n = number of generations). Hence, G=t/n is the equation from which calculations of generation time (below) derive.
3. *Stationary Phase:* Exponential growth cannot be continued forever in a batch culture (*e.g.*, a closed system such as a test tube or flask). Population growth is limited by one of three factors: 1. exhaustion of available nutrients; 2. accumulation of inhibitory metabolites or end products; 3. exhaustion of space, in this case called a lack of "biological space".

   During the stationary phase, if viable cells are being counted, it cannot be determined whether some cells are dying and an equal number of cells are dividing, or the population of cells has simply stopped growing and dividing. The stationary phase, like the lag phase, is not necessarily a period of quiescence. Bacteria that produce secondary metabolites, such as antibiotics, do so during the stationary phase of the growth cycle (Secondary metabolites are defined as metabolites produced after the active stage of growth). It is during the stationary phase that spore-forming bacteria have to induce or unmask the activity of dozens of genes that may be involved in sporulation process.
4. *Death Phase*: If incubation continues after the population reaches stationary phase, a death phase follows, in which the viable cell population declines. (Note, if counting by turbidimetric measurements or microscopic counts, the death phase cannot be observed.). During the death phase, the number of viable cells decreases geometrically (exponentially), essentially the reverse of growth during the log phase.

GROWTH RATE AND GENERATION TIME

As mentioned above, bacterial growth rates during the phase of exponential growth, under standard nutritional conditions (culture medium, temperature, pH, etc.), define the bacterium's generation time. Generation times for bacteria vary from about 12 minutes to 24 hours or more. The generation time for *E. coli* in the laboratory is 15-20 minutes, but in the intestinal tract, the coliform's generation time is estimated to be 12-24 hours. For most known bacteria that can be cultured, generation times range from about 15 minutes to 1 hour. Symbionts such as *Rhizobium* tend to have longer generation times. Many lithotrophs, such as the nitrifying bacteria, also have long generation times. Some bacteria that are pathogens, such as *Mycobacterium tuberculosis* and *Treponema pallidum*, have especially long generation times, and this is thought to be an advantage in their virulence.

Table. Generation Times for Some Common Bacteria Under Optimal Conditions of Growth.

Bacterium	Medium	Generation Time (Minutes)
Escherichia coli	Glucose-salts	17
Bacillus megaterium	Sucrose-salts	25
Streptococcus lactis	Milk	26
Streptococcus lactis	Lactose broth	48
Staphylococcus aureus	Heart infusion broth	27-30
Lactobacillus acidophilus	Milk	66-87
Rhizobium japonicum	Mannitol-salts-yeast extract	344-461
Mycobacterium tuberculosis	Synthetic	792-932
Treponema pallidum	Rabbit testes	1980

CALCULATION OF GENERATION TIME

When growing exponentially by binary fission, the increase in a bacterial population is by geometric progression. If we start with one cell, when it divides, there are 2 cells in the first generation, 4 cells in the second generation, 8 cells in the third generation, and so on.

The generation time is the time interval required for the cells (or population) to divide:

G (generation time) = (time, in minutes or hours)/n(number of generations)

$G = t/n$

t – time interval in hours or minutes

B = number of bacteria at the beginning of a time interval

b = number of bacteria at the end of the time interval

n = number of generations (number of times the cell population doubles during the time interval)

$b = B \times 2^n$ (This equation is an expression of growth by binary fission)

Solve for n:

$$\log b = \log B + n\log 2$$

$$n = \frac{\log b - \log B}{\log 2}$$

$$n = \frac{\log b - \log B}{.301}$$

$$n = 3.3 \log b/B$$

$$G = t/n$$

Solve for G

$$G = \frac{t}{3.3 \log b / B}$$

Example: What is the generation time of a bacterial population that increases from 10,000 cells to 10,000,000 cells in four hours of growth?

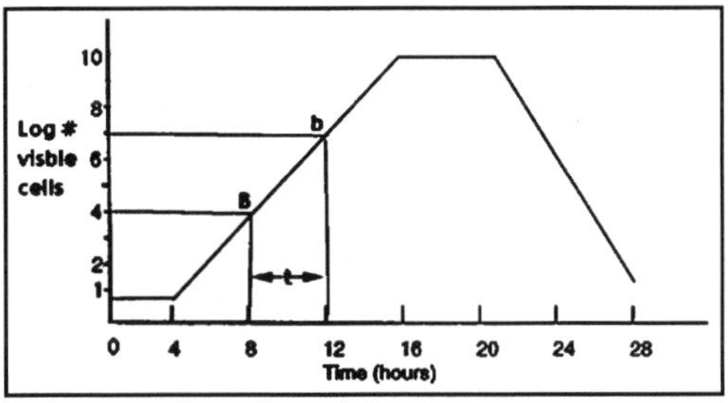

$$G = \frac{t}{3.3 \log b / B}$$

$$G = \frac{240 \text{ minutes}}{3.3 \log 10^7 / 10^4}$$

$$G = \frac{240 \text{ minutes}}{3.3 \times 3}$$

$$G = 24 \text{ minutes}$$

CONTINUOUS CULTURE OF BACTERIA

The cultures so far discussed for growth of bacterial populations are called batch cultures. Since the nutrients are not renewed, exponential growth is limited to a few generations. Bacterial cultures can be maintained in a state of exponential growth over long periods of time using a system of continuous culture, designed to relieve the conditions that stop exponential growth in batch cultures. Continuous culture, in a device called a chemostat, can be used to maintain a bacterial population at a constant density, a situation that is, in many ways, more similar to bacterial growth in natural environments.

In a chemostat, the growth chamber is connected to a reservoir of sterile medium. Once growth is initiated, fresh medium is continuously supplied from the reservoir. The volume of fluid in the growth chamber is maintained at a constant level by some sort of overflow drain. Fresh medium is allowed to enter into the growth chamber at a rate that limits the growth of the bacteria. The bacteria grow (cells are formed) at the same rate that bacterial cells (and spent medium) are removed by the overflow. The rate of addition of the fresh medium determines the rate of growth because the fresh medium always contains a limiting amount of an essential nutrient. Thus, the chemostat relieves the insufficiency of nutrients, the accumulation of toxic substances, and the accumulation of excess cells in the culture, which are the parameters that initiate the stationary phase of the growth cycle. The bacterial culture can be grown and maintained at relatively constant conditions, depending on the flow rate of the nutrients.

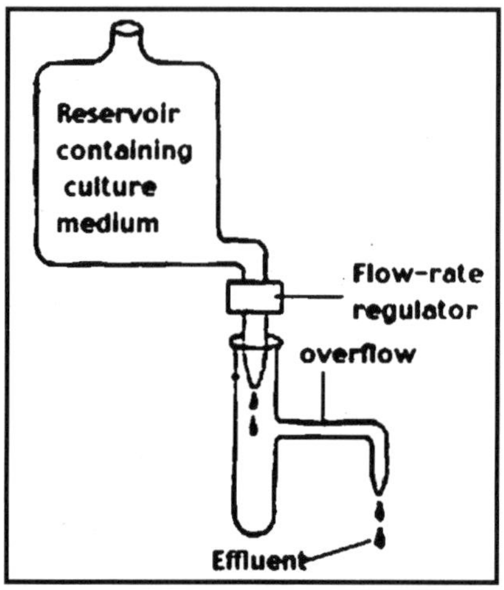

Fig. Schematic Diagram of a Chemostat, a Device for the Continuous Culture of Bacteria. The Chemostat Relieves the Environmental Conditions that Restrict Growth by Continuously Supplying Nutrients to Cells and Removing Waste Substances and Spent Cells from the Culture Medium.

SYNCHRONOUS GROWTH OF BACTERIA

Studying the growth of bacterial populations in batch or continuous cultures does not permit any conclusions about the growth behaviour of individual cells, because the distribution of cell size (and hence cell age) among the members of the population is completely random. Information about the growth behaviour of individual bacteria can, however, be obtained by the study of synchronous cultures. Synchronised cultures must be composed of cells which are all at the same stage of the bacterial cell cycle. Measurements made on synchronised cultures are equivalent to measurements made on individual cells.

A number of clever techniques have been devised to obtain bacterial populations at the same stage in the cell cycle. Some techniques involve manipulation of environmental parameters which induces the population to start or stop growth at the same point in the cell cycle, while others are physical methods for selection of cells that have just completed the process of binary fission. Theoretically, the smallest cells in a bacterial population are those that have just completed the process of cell division.

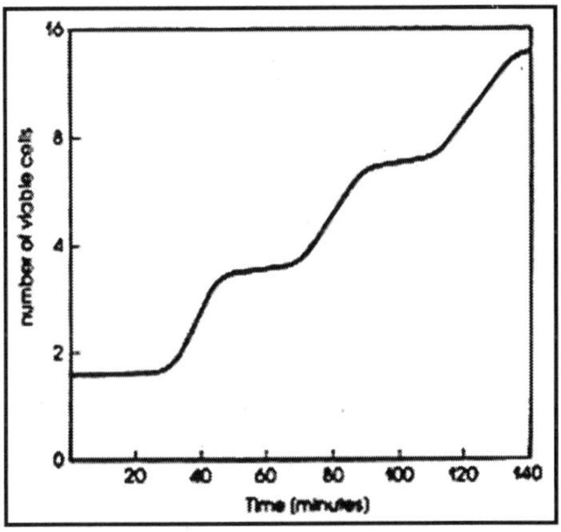

Fig. The Synchronous Growth of a Bacterial Population. By Careful Selection of Cells that have Just Divided, A Bacterial Population can be Synchronised in the Bacterial Cell Division Cycle. Synchrony can be Maintained for Only a Few Generations.

BACTERIA AND ARCHAEA

The most significant effect that the procaryotes, bacteria and archaea, have on their environment is their underlying ability to recycle the essential elements that make up cells. The earth is a closed system with limited amounts of certain elements in forms that are utilised by cells.

These element are generally acted upon first by microbes to assimilate them into living matter. The total biomass of microbial cells in the biosphere, their metabolic diversity, and their persistence in all habitats that support life, guarantee that microbes will play crucial roles in the transformations and recycling of these elements among all forms of life.

The table below lists the major elements that make up a typical procaryotic cell (in this case, *E. coli*). As expected, over 90 percent of the elemental analysis consists of carbon, hydrogen, oxygen, nitrogen, phosphorus and sulphur. These are the elements that become combined to form all the biochemicals and macromolecules that comprise living systems. C, H, O, N, P and S are the constituents of organic material (An organic compound is a chemical that contains a carbon to hydrogen bond.

Organic compounds on earth are evidence of life. Organic compounds may be symbolised as CH_2O, which is the empirical formula for a sugar such as glucose.) H and O are the constituents of water (H_2O), that makes up over 95 percent of the cell composition. Calcium (Ca^{++}), iron (Fe^{++}), magnesium (Mg^{++}) and potassium (K^+) are present as inorganic salts in the cytoplasm of cells.

Table. Major Elements, their Sources and Functions in Cells.

Element	Per cent of dry weight	Source	Function
Carbon	50	Organic compounds or CO_2	Main constituent of cellular material
Oxygen	20	H_2O, organic compounds, CO_2, and O_2	Constituent of cell material and cell water; O_2 is electron acceptor in aerobic respiration
Nitrogen	14	NH_3, NO_3, organic compounds, N_2	Constituent of amino acids, nucleic acids nucleotides, and coenzymes
Hydrogen	8	H_2O, organic compounds, H_2	Main constituent of organic compounds and cell water
Phosphorus	3	inorganic phosphates (PO_4)	Constituent of nucleic acids, nucleotides, phospholipids, LPS, teichoic acids
Sulfur	1	SO_4, H_2S, S, organic sulfur compounds	Constituent of cysteine, methionine, glutathione, several coenzymes
Potassium	1	Potassium salts	Main cellular inorganic cation and cofactor for certain enzymes
Magnesium	0.5	Magnesium salts	Inorganic cellular cation, cofactor for certain enzymatic reactions
Calcium	0.5	Calcium salts	Inorganic cellular cation, cofactor for certain enzymes and a component of endospores
Iron	0.2	Iron salts	Component of cytochromes and certain nonheme iron-proteins and a cofactor for some enzymatic reactions

The table ignores the occurrence of "trace elements" in cells. Trace elements are metal ions required in cellular nutrition in such small amounts that it is difficult to determine or demonstrate their presence in cells. The usual metals that qualify as trace elements are Mn^{++}, Co^{++}, Zn^{++}, Cu^{++} and Mo^{++}. Trace elements are usually built into vitamins and enzymes. For example, vitamin B_{12} contains cobalt (Co^{++}) and the bacterial nitrogenase enzyme contains molybdenum (Mo^{++}).

MICROBES AND THE CYCLES OF ELEMENTS(H)

Of course, all living organisms play a role in the cycles of elements, but for the most part, it is the procaryotes that play major and sometimes unique roles. Herein, we discuss total microbial contribution to the cycles of the major elements, but major emphasis is placed on procaryotes.

The fungi (molds and yeasts). The molds are aerobic organisms that utilise organic compounds for growth. They play an important role in decomposition or biodegradation of organic matter, particularly in soil. Yeast can grow anaerobically (without oxygen) through the process of fermentation. They play a role in fermentations in environments high in sugar. The prominent role of fungi in the environment is in the carbon cycle, during the process of decomposition, especially in the soil.

The algae are also an important part of the carbon cycle. They are the predominant photosynthetic organisms in many aquatic environments. The algae are autotrophs, which means they use carbon dioxide (CO_2) as a source of carbon for growth. Hence they convert atmospheric CO_2 into organic material (*i.e.*, algal cells). Algae also play a role in the oxygen (O_2) cycle since their style of photosynthesis, similar to plants, produces O_2 in the atmosphere. The cyanobacteria are a group of procaryotic microbes, as prevalent as algae, that have this type of metabolism. Photosynthetic algae and cyanobacteria can be found in most environments where there is moisture and light. They are a major component of marine plankton which the basis of the food chain in the oceans.

Protozoans are heterotrophic organisms that have to catch or trap their own food. Therefore, they have developed elaborate mechanisms for movement and acquiring organic food which they can digest. Their food usually turns out to be bacterial cells, so one might argue that they are ecological predators that keep bacterial populations under control in soil, aquatic environments, intestinal tracts of animals, and many other environments.

The procaryotic bacteria and archaea, as a result of their diversity and unique types of metabolism, are involved in the cycles of virtually all essential elements. In two cases, methanogenesis (conversion of carbon dioxide into methane) and nitrogen fixation (conversion of nitrogen in the atmosphere into biological nitrogen) are unique to procaryotes and earns them their "essential role" in the carbon and nitrogen cycles.

There are other metabolic processes that are unique, or nearly so, in the procaryotes that bear significantly on the cycles of elements. For example, procaryotes called lithotrophs use inorganic compounds like ammonia and hydrogen sulfide as a source of energy, and others called anaerobic respirers use nitrate (NO_3) or sulfate (SO_4) in the place of oxygen, so they can respire without air. Most of the archaea are lithotrophs that use hydrogen (H_2) or hydrogen sulfide (H_2S) as a source of energy, while many soil bacteria are anaerobic respirers that can use their efficient respiratory metabolism in the absence of O_2. The basic processes of heterotrophy are spread throughout the bacteria. Most of the bacteria in the soil and water, and in associations with animals and plants, are heterotrophs. Heterotrophy means living off of dead organic matter, usually by some means of respiration (same as animals) or fermentation (same as yeast or lactic acid bacteria). Bacterial heterotrophs in the carbon chain are important in the processes of biodegradation and decomposition under aerobic and anaerobic conditions.

In bacteria, there is a unique type of photosynthesis that does not use H_2O or produce O_2 which impacts on the carbon and sulphur cycles.

Meanwhile, the cyanobacteria (mentioned above) fix CO_2 and produce O_2 during photosynthesis, and they make a very large contribution to the carbon and oxygen cycles.

The list of examples of microbial involvement in the cycles of elements that make up living systems is endless, and probably every microbe in the web is involved in an intimate and unique way.

THE OXYGEN CYCLE

Basically, O_2 is derived from the photolysis of H_2O during plant (oxygenic) photosynthesis. Two major groups of microorganisms are involved in this process, the eucaryotic algae, and the procaryotic cyanobacteria (formerly known as "blue-green algae"). The cyanobacteria and algae are the source of much of the O_2 in the earth's atmosphere. Of course, plants account for some O_2 production as well, but the microbes predominate in marine habitats which cover the majority of the planet.

Since most aerobic organisms need the O_2 that results from plant photosynthesis, this establishes a relationship between plant photosynthesis and aerobic respiration, two prominent types of metabolism on earth. Photosynthesis produces O_2 needed for aerobic respiration.

Respiration produces CO_2 needed for autotrophic growth:

$CO_2 + H_2O \rightarrow CH_2O$ (organic material) $+ O_2$ plant (oxygenic) photosynthesis

$CH_2O + O_2 \rightarrow CO_2 + H_2O$ aerobic respiration

Since these photosynthetic microbes are also autotrophic (meaning they convert CO_2 to organic material during growth) they have a similar impact on the carbon cycle.

THE CARBON CYCLE

Carbon is the backbone of all organic molecules and is the most prevalent element in cellular (organic) material. In its most oxidised form, CO_2, it can be viewed as an "inorganic" molecule (no C - H bond). Autotrophs, which include plants, algae, photosynthetic bacteria, lithotrophs, and methanogens, use CO_2 as a sole source of carbon for growth, which reduces the molecule to organic cell material (CH_2O).

Heterotrophs require organic carbon for growth, and ultimately convert it back to CO_2. Thus, a relationship between autotrophs and heterotrophs is established wherein autotrophs fix carbon needed by heterotrophs, and heterotrophs produce CO_2 used by the autotrophs.

$CO_2 + H_2O \rightarrow CH_2O$ (organic material) autotrophy

$CH_2O + O_2 \rightarrow CO_2 + H_2O$ heterotrophy

Since CO_2 is the most prevalent greenhouse gas in the atmosphere, it isn't good if these two equations to get out of balance (*i.e.*, heterotrophy predominating over autotrophy, as when rain forests are destroyed and replaced with cattle). Autotrophs are referred to as primary producers at the "bottom of the food chain" because they convert carbon to a form required by heterotrophs. Among procaryotes, the cyanobacteria, the lithotrophs and the methanogens are a formidable biomass of autotrophs that account for a corresponding amount of CO_2 fixation in the global carbon cycle.

The lithotrophic bacteria and archaea that oxidise reduced N and S compounds and play important roles in the natural cycles of N and S (discussed below), are virtually all autotrophs. The prevalence of these organisms in sulphur-rich environments (marine sediments, thermal vents, hot springs, endosymbionts, etc., may indicate an unappreciated role of these procaryotes as primary producers of organic carbon on earth. The methanogens play a dual role in the carbon cycle.

These archaea are inhabitants of virtually all anaerobic environments in nature where CO_2 and H_2 (hydrogen gas) occur. They use CO_2 in their metabolism in two distinct ways. About 5 percent of CO_2 taken up is reduced to cell material during autotrophic growth; the remaining 95 percent is reduced to CH_4 (methane gas) during a unique process of generating cellular energy. Hence, methane accumulates in rocks as fossil fuel ("natural gas"), in the rumen of cows and guts of termites, in sediments, swamps, landfills and sewage digesters. Since CH_4 is the second-most prevalent of the greenhouse gases, it is best to discourage processes that lead to its accumulation in the atmosphere.

$CO_2 + H_2 \rightarrow CH_2O$ (cell material) $+ CH_4$ methanogenesis

Under aerobic conditions, methane and its derivatives (methanol, formaldehyde, etc.) can be oxidised as energy sources by bacteria called methylotrophs. Metabolically this is a version of decomposition or biodegradation during the carbon cycle which is discussed below. Biodegradation is the process in the carbon cycle for which microbes get most credit (or blame). Biodegradation is the decomposition of organic material (CH_2O) back to CO_2 + H_2O and H_2.

In soil habitats, the fungi play a significant role in biodegradation, but the procaryotes are equally important. The typical decomposition scenario involves the initial degradation of biopolymers (cellulose, lignin, proteins, polysaccharides) by extracellular enzymes, followed by oxidation (fermentation or respiration) of the monomeric subunits. The ultimate end products are CO_2, H_2O and H_2, perhaps some NH_3 (ammonia) and sulfide (H_2S), depending on how one views the overall process.

These products are scarfed up by lithotrophs and autotrophs for recycling. Procaryotes which play an important role in biodegradation in nature include the actinomycetes, clostridia, bacilli, arthrobacters and pseudomonads.

OVERALL PROCESS OF BIODEGRADATION (DECOMPOSITION)

polymers (*e.g.*, cellulose) → monomers (*e.g.*, glucose) depolymerisation

monomers → fatty acids (*e.g.*, lactic acid, acetic acid, propionic acid) $+ CO_2 + H_2$ fermentation

monomers $+ O_2 \rightarrow CO_2 + H_2O$ aerobic respiration

The importance of microbes in biodegradation is embodied in the adage that "there is no known natural compound that cannot be degraded by some microorganism." The proof of the adage is that we aren't up to our ears in whatever it is that couldn't be degraded in the last 3.5 billion years. Actually, we are up to our ears in cellulose and lignin, which is better than concrete, and some places are getting up to their ears in teflon, plastic, styrofoam, insecticides, pesticides and poisons that are degraded slowly by microbes, or not at all.

The figure above mostly ignores the role of methanogenesis in the carbon cycle. Since methanogens have the potential to remove CO_2 from the atmosphere, converting it to cell material and CH_4, these procaryotes not only influence the carbon cycle, but their metabolism also affects the concentration of major greenhouse gases in earth's atmosphere.

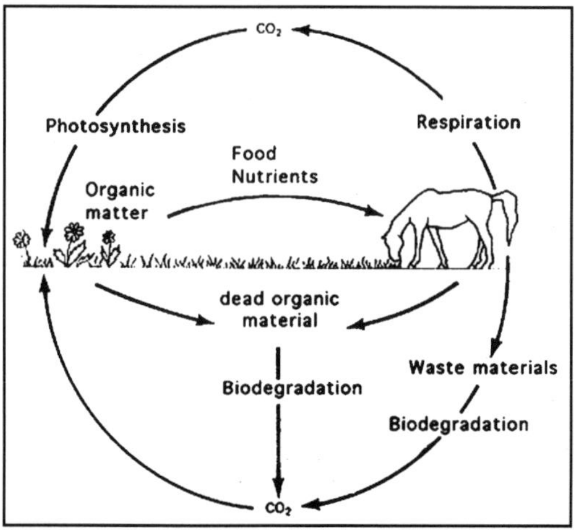

Fig. The Carbon Cycle. Organic Matter (CH_2O) Derived from Photosynthesis (Plants, Algae and Cyanobacteria) Provides Nutrition for Heterotrophs (*e.g.* Animals and Associated Bacteria), which Convert it Back to CO_2. Organic Wastes, as well as Dead Organic Matter in the Soil and Water, are Ultimately Broken Down to CO_2 by Microbial Processes of Biodegradation.

During methanogenesis involving CO_2 reduction, the stoichiometry is $4H_2 + CO_2 \longrightarrow CH_4 + 2\ H_2O$, so one mole of a greenhouse gas is exchanged for another. But methane is about 15 times more potent than is CO_2 in terms of heat absorption capability on a per-molecule basis, so the net effect is a functional increase in heat absorption by the atmosphere. Remember also that in most natural environments, around two-thirds of the methane is produced by aceticlastic methanogenesis ($CH_3COOH \longrightarrow CH_4 + CO_2$) - an even less welcome situation, as BOTH products are greenhouse gases.

Even though methane concentrations in the atmosphere are two orders of magnitude below those of CO_2, methane is thought to account for about 15 per cent of the anthropogenic climate forcing, compared to about 60 per cent from CO_2.

Most of the rest of the contribution is from nitrous oxide (N_2O, a respiratory denitrification product that has something like 300 times the heat absorbing capacity as CO_2) and the old chlorofluorocarbons (CFCs), even stronger heat absorbers yet, but more famous and dangerous as stratospheric ozone-depleters."

THE NITROGEN CYCLE

The nitrogen cycle is the most complex of the cycles of elements that make up biological systems. This is due to the importance and prevalence of N in cellular metabolism, the diversity of types of nitrogen metabolism, and the existence of the element in so many forms. Procaryotes are essentially involved in the biological nitrogen cycle in three unique processes.

- *Nitrogen Fixation*: This process converts N_2 in the atmosphere into NH_3 (ammonia), which is assimilated into amino acids and proteins. Nitrogen fixation occurs in many free-living bacteria such as clostridia, azotobacters and cyanobacteria, and in symbiotic bacteria such as *Rhizobium* and *Frankia,* which associate with plant roots to form characteristic nodules. Biological nitrogen fixation is the most important way that N_2 from the air enters into biological systems.

$$N_2 \rightarrow 2\ NH_3 \text{ nitrogen fixation}$$

- *Anaerobic Respiration*: This relates to the use of oxidised forms of nitrogen (NO_3 and NO_2) as final electron acceptors for respiration. Anaerobic respirers such as *Bacillus* and *Pseudomonads are* common soil inhabitants that will use nitrate (NO_3) as an electron acceptor. NO_3 is reduced to NO_2 (nitrite) and then to a gaseous form of nitrogen such as N_2 or N_2O (nitrous oxide). The process is called denitrification. (A related process conducted by some *Bacillus* species, called dissimilatory nitrate reduction reduces NO_3 to ammonia (NH_3), but this is not considered denitrification.) Denitrifying bacteria are typically facultative microbes that respire whenever oxygen is available by aerobic respiration. If O_2 is unavailable for respiration, they will turn to the alternative anaerobic respiration which uses NO_3. Since NO_3 is a common and expensive form of fertilizer in soils, denitrification may not be so good for agriculture, and one rationale for tilling the soil is to keep it aerobic, thereby preserving nitrate fertilizer in the soil.

$$NO_3 \rightarrow NO_2 \rightarrow N_2 \text{ denitrification}$$

The overall reactions of denitrification shown above proceed through the formation of nitrous oxide (N_2O). A recent article by Wunsch an Zumft in Journal of Bacteriology, vol. 187 (2005), sheds new light on the process of denitrification. N_2O is a bacterial metabolite in the REVERSAL of Nitrogen fixation. The anthropogenic atmospheric increase of N_2O is a cause for concern, as noted above (as a greenhouse gas, N_2O has 300 times the heat absorbing capacity as CO_2). Denitrifying bacteria respire using N_2O as an electron acceptor yielding N_2 and the thereby provide a sink for N_2O. This article provides new insight into this process by identifying a membrane-bound protein in denitrifying bacteria called NosR, that is necessary for the expression of N_2O reductase from the *nosZ* gene. The NosR protein has redox centers positioned on opposite sides of the cytoplasmic membrane, which allows it to sustain whole-cell N_2O respiration by acting on N_2O reductase. Nitrification is a form of lithotrophic metabolism that is chemically the opposite of denitrification. Nitrifying bacteria such as *Nitrosomonas* utilise NH_3 as an energy source, oxidising it to NO_2, while *Nitrobacter* will oxidise NO_2 to NO_3. Nitrifying bacteria generally occur in aquatic environments and their significance in soil fertility and the global nitrogen cycle is not well understood.

The Overall process of Nitrification:

$NH_3 \rightarrow NO_2$ (*Nitrosomonas*)

$NO_2 \rightarrow NO_3$ (*Nitrobacter*)

A final important aspect of the nitrogen cycle that involves procaryotes, though not exclusively, is decomposition of nitrogen-containing compounds. Most organic nitrogen (in protein, for example) yields ammonia (NH_3) during the process of deamination. Fungi are involved in decomposition, as well.

Plants, animals and protista, as well as the procaryotes, complete the nitrogen cycle during the uptake of the element for their own nutrition. Nitrogen assimilation is usually in the form of nitrate, an amino group, or ammonia.

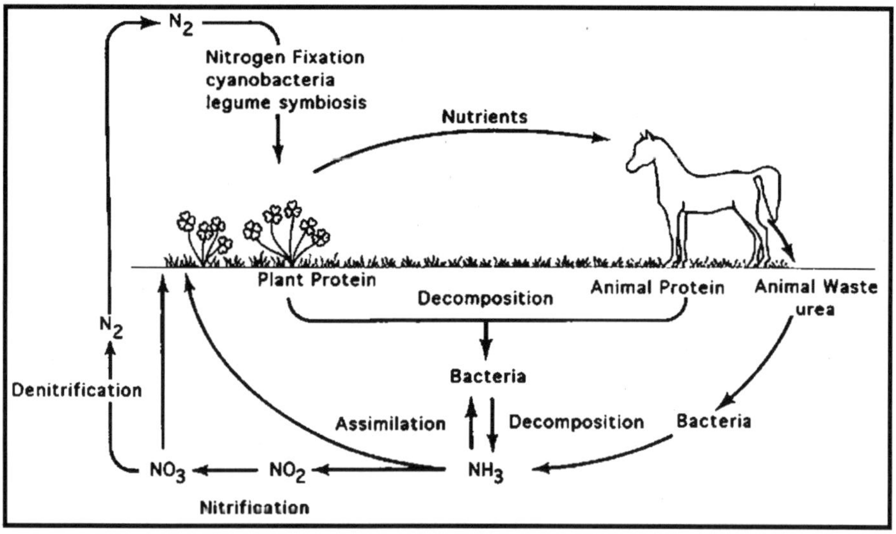

Fig. The Nitrogen Cycle.

THE SULPHUR CYCLE

Sulphur is a component of a couple of vitamins and essential metabolites and it occurs in two amino acids, cysteine and methionine. In spite of its paucity in cells, it is an absolutely essential element for living systems. Like nitrogen and carbon, the microbes can transform sulphur from its most oxidised form (sulfate or SO_4) to its most reduced state (sulfide or H_2S). The sulphur cycle, in particular, involves some unique groups of procaryotes and procaryotic processes.

Two unrelated groups of procaryotes oxidise H_2S to S and S to SO_4. The first is the anoxygenic photosynthetic purple and green sulphur bacteria that oxidise H_2S as a source of electrons for cyclic photophosphorylation. The second is the "colourless sulphur bacteria" (now a misnomer because the group contains many Archaea) which oxidise H_2S and S as sources of energy. In either case, the organisms can usually mediate the complete oxidation of H_2S to SO_4.

$$H_2S \rightarrow S \rightarrow SO_4$$

litho or phototrophic sulphur oxidation

Sulphur-oxidising procaryotes are frequently thermophiles found in hot (volcanic) springs and near deep sea thermal vents that are rich in H_2S. They may be acidophiles, as well, since they acidify their own environment by the production of sulphuric acid. Since SO_4 and S may be used as electron acceptors for respiration, sulfate reducing bacteria produce H_2S during a process of anaerobic respiration analogous to denitrification. The use of SO_4 as an electron acceptor is an obligatory process that takes place only in anaerobic environments. The process results in the distinctive odour of H_2S in anaerobic bogs, soils and sediments where it occurs. Sulphur is assimilated by bacteria and plants as SO_4 for use and reduction to sulfide. Animals and bacteria can remove the sulfide group from proteins as a source of S during decomposition. These processes complete the sulphur cycle.

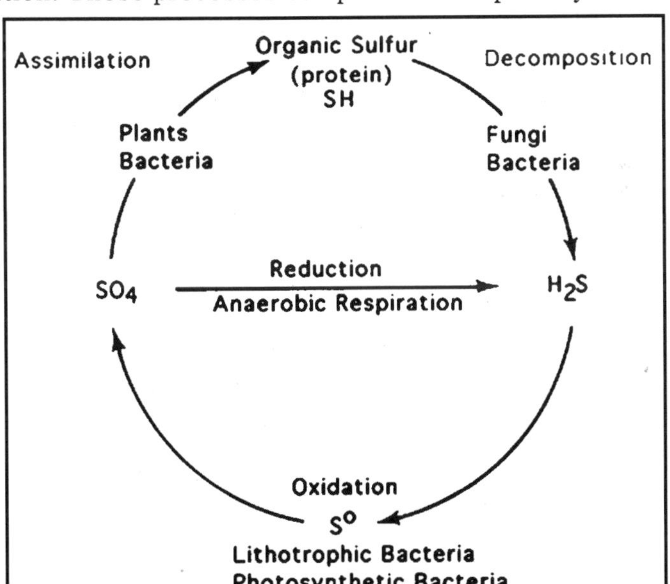

Fig.. The Sulphur Cycle.

THE PHOSPHORUS CYCLE

The phosphorus cycle is comparatively simple. Inorganic phosphate exists in only one form. It is interconverted from an inorganic to an organic form and back again, and there is no gaseous intermediate. Phosphorus is an essential element in biological systems because it is a constituent of nucleic acids, (DNA and RNA) and it occurs in the phospholipids of cell membranes. Phosphate is also a constituent of ADP and ATP which are universally involved in energy exchange in biological systems.

Dissolved phosphate (PO_4) inevitably ends up in the oceans. It is returned to land by shore animals and birds that feed on phosphorus containing sea creatures and then deposit their feces on land. Dissolved PO_4 is also returned to land by a geological process, the uplift of ocean floors to form land masses, but the process is very slow.

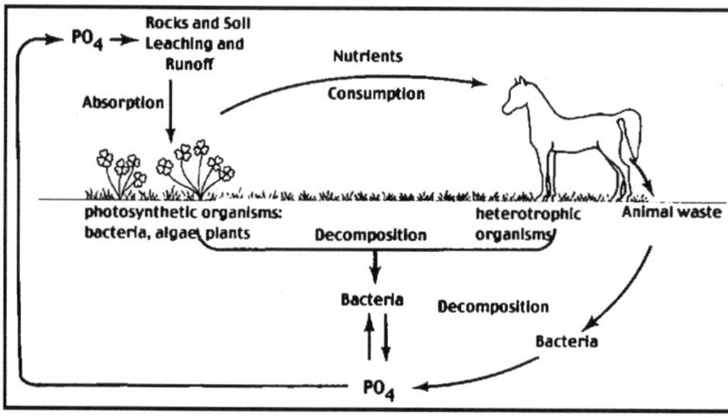

Fig. The Phosphorus Cycle. Plants, Algae and Photosynthetic Bacteria can Absorb Phosphate (PO_4) Dissolved in Water, or if it Washes Out of Rocks and Soils. They Incorporate the PO_4 into Various Organic Forms, Including such Molecules as DNA, RNA, ATP, and Phospholipid. The Plants are Consumed by Animals wherein the Organic Phosphate in the Plant Becomes Organic Phosphate in the Animal and in the Bacteria that live with the Animal. Animal Waste Returns Inorganic PO_4 to the Environment and Also Organic Phosphate in the Form of Microbial Cells. Dead Plants and Animals, as well as Animal Waste, are Decomposed by Microbes in the Soil. The Phosphate Eventually is Mineralised to the Soluble PO_4 Form in Water and Soil, to be taken Up Again by Photosynthetic Organisms.

ECOLOGY OF A STRATIFIED LAKE

The surface of the lake is well-lighted by the sun and is aerobic. The bottom of the lake and its sediments are dark and anaerobic. Generally there is less O_2 and less light as the water column is penetrated from the surface.

Assuming that the nutrient supply is stable and there is no mixing between layers of lake water, we should, for the time being, have a stable ecosystem with recycling of essential elements among the living systems. Here is how it would work.

At the surface, light and O_2 are plentiful, CO_2 is fixed and O_2 is produced. Photosynthetic plants, algae and cyanobacteria produce O_2, cyanobacteria can even fix N_2; aerobic bacteria, insects, animals and plants live here.

At the bottom of the lake and in the sediments, conditions are dark and anaerobic. Fermentative bacteria produce fatty acids, H_2 and CO_2, which are used by methanogens to produce CH_4. Anaerobic respiring bacteria use NO_3

and SO_4 as electron acceptors, producing NH_3 and H_2S. Several soluble gases are in the water: H_2, CO_2, CH_4, NH_3 and H_2S. The biological activity at the surface of the lake and at the bottom of the lake may have a lot to do with what will be going on in the middle of the water column, especially near the interface of the aerobic and anaerobic zones.

This area, called the thermocline, is biologically very active. Bacterial photosynthesis, which is anaerobic, occurs here, using longer wave lengths of light that will penetrate the water column and are not absorbed by all the plant chlorophyll above. The methanotrophs will stay just within the aerobic area taking up the CH_4 from the sediments as a carbon source, and returning it as CO_2. Lithotrophic nitrogen and sulphur utilising bacteria do something analogous: they are aerobes that use NH_3 and H_2S from the sediments, returning them to NO_3 and SO_4.

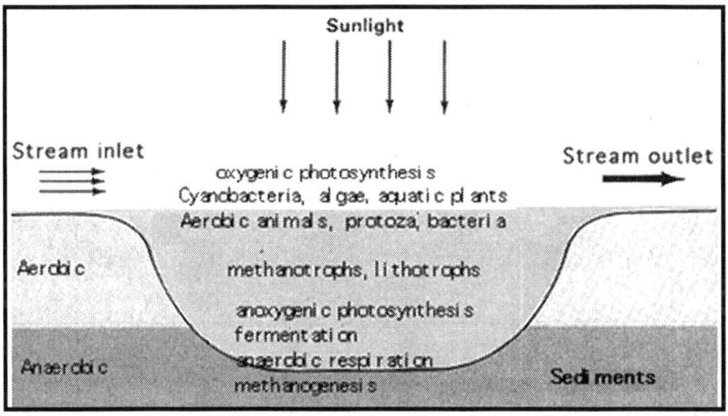

Fig. Ecology of a Stratified Lake.

IMPORTANT GROUPS OF PROCARYOTES

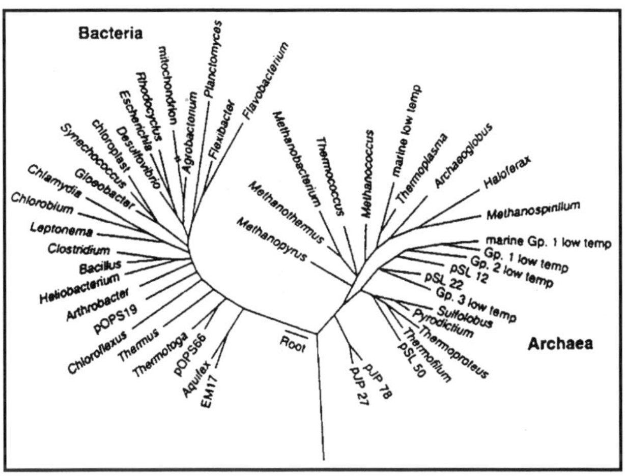

The procaryotes (or prokaryotes) consist of millions of genetically-distinct unicellular organisms. They lack structural diversity but have a wide range of genetic and physiological diversity. Sometimes, particular physiological traits unify and distinguish particular groups of procaryotes to microbiologists. In Bergey's Manual of Determinative Bacteriology (1994), the identifiable groups of procaryotes are assembled based on easily-observed phenotypic characteristics such as Gram stain, morphology (rods, cocci, etc.), motility, structural features (*e.g.*, spores, filaments, sheaths, appendages, etc.), and on distinguishing physiological features (*e.g.*, photosynthesis, anaerobiasis, methanogenesis, lithotrophy, etc.).

Nowadays, this type of artificial classification scheme has been abandoned in favour of hierarchical taxonomic schemes based on comparative genetic analysis of the nucleotide sequences of the small subunit ribosomal RNA that is contained in all cellular organisms.

In the Second edition of Bergey's Manual of Systematic Bacteriology (2001-2004), as well as the current (3rd) edition of The Prokaryotes, phylogeny dominates the classification schemes. Such an approach generates the Phylogenetic Tree of Life (below) that lands the procaryotes in two Domains, Archaea and Bacteria.

At a taxonomic level, organisms at the tips of the archaeal branches represent Orders; the tips of the bacterial branches are Phyla. More information on the taxonomy, phylogeny and classification of procaryotes is given in the references at the bottom of this page. Also, an excellent article online that integrates phylogeny with classification of procaryotes is Classification and Phylogeny by Gary Olsen.

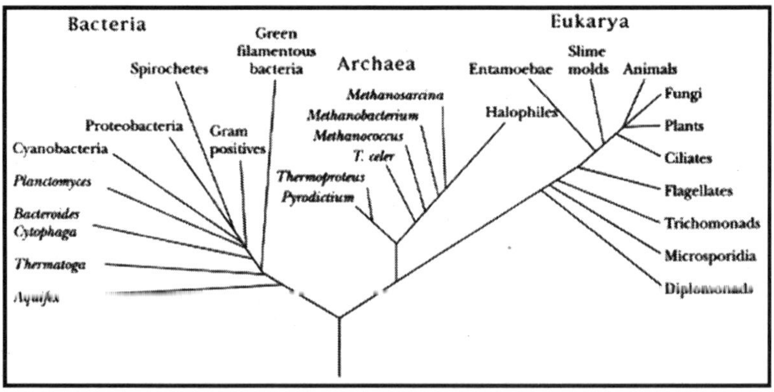

The Phylogenetic Tree of Life based on comparative ssrRNA Sequencing. The Tree shows the procaryotes in two Domains, Archaea and Bacteria. At a taxonomic level, most organisms at the tips of the Archaeal branches represent a unique Order; most organisms a the tips of the bacterial branches are classified into a unique Phylum.

On the Archaeal limb, the three physiological groups are evident in the names: "thermo" and "pyro" for the extreme thermophiles; "methano" for the methanogens; and "halophiles" for the extreme halophiles. The most important, best known, and diverse groups (phyla) branching off of the Bacterial limb are the Cyanobacteria, Proteobacteria and Gram positives. In the ensuing description of "important groups of procaryotes", groups of organisms are placed under artificial headings based on common structural, biochemical or ecophysiological properties. This does not imply close genetic relatedness among different genera in a group. Sometimes, all of the members of a group do share a close genetic relatedness; in other cases, members of a group are genetically-unrelated, even to an extent that is greater than exists among all members of the Eucaryotic domain. Also herein, some procaryotes are placed in more than one group, and occasionally some groups consist of both Archaea and Bacteria.

ARCHAEA

On the basis of ssrRNA analysis, the Archaea consist of three phylogenetically-distinct groups: Crenarchaeota, Euryarchaeota and Korarchaeota. However, for the Korarchaeota, only their nucleic acids have been detected, and no organisms have been isolated or cultured.

Based on their physiology, the Archaea can be organised into three types:

1. Methanogens (procaryotes that produce methane);
2. Extreme halophiles (procaryotes that live at very high concentrations of salt (NaCl); and
3. Extreme (hyper) thermophiles (procaryotes that live at very high temperatures).

In addition to the unifying archaeal features that distinguish them from Bacteria (*i.e.*, no murein in cell wall, ether-linked membrane lipids, etc.), the Archaea exhibit other unique structural or biochemical attributes which adapt them to their particular habitats.

The Crenarchaeota consist mainly of hyperthermophilic sulphur-dependent procaryotes and the Euryarchaeota contains the methanogens and extreme halophiles. ssrRNAs of the Korarchaeota have been obtained from hyperthermophilic environments similar to those inhabited by Crenarchaeota. None of the Korarchaeota have been cultured in the laboratory, although information about them can be inferred from their genome structure. Methanogens are obligate anaerobes that will not tolerate even brief exposure to air (O_2). Anaerobic environments are plentiful, however, and include marine and fresh-water sediments, bogs and deep soils, intestinal tracts of animals, and sewage treatment facilities.

Methanogens have an incredible type of metabolism that can use H_2 as an energy source and CO_2 as a carbon source for growth. In the process of making cell material from H_2 and CO_2, the methanogens produce methane (CH_4) in a unique energy-generating process. The end product (methane gas) accumulates in their environment. Methanogen metabolism created most the natural gas (fossil fuel) reserves that are tapped as energy sources for domestic or industrial use. Methanogens are normal inhabitants of the rumen (fore-stomach) of cows and other ruminant animals.

A cow belches about 50 litres of methane a day during the process of eructation (chewing the cud). Methane is a significant greenhouse gas and is accumulating in the atmosphere at an alarming rate.

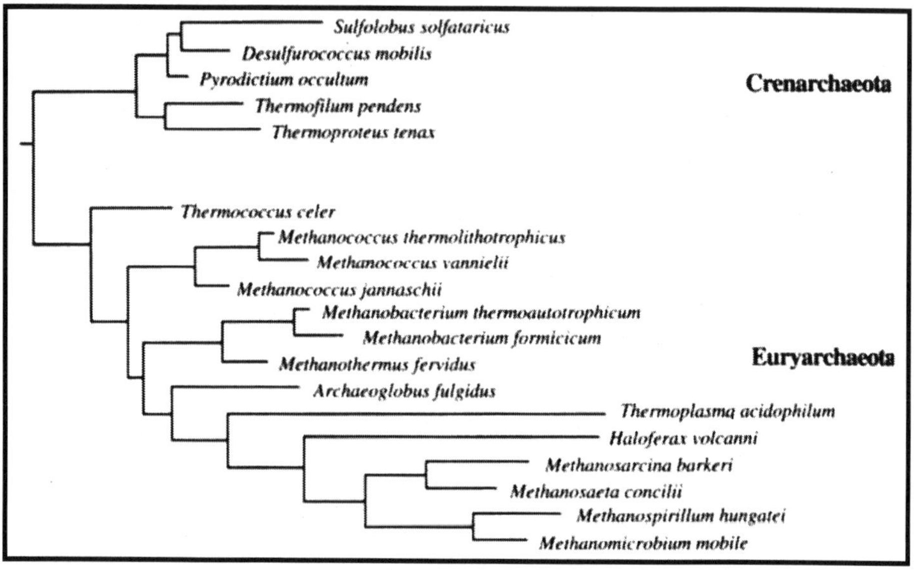

Fig. Phylogenetic Tree of Archaea.

When rain forests are destroyed and replaced by cows, it is "double-hit" on the greenhouse:

(1) Less CO_2 is taken up due to removal of the autotrophic green plants;

(2) Additional CO_2 and CH_4 are produced as gases by the combined metabolism of the animal and symbiotic methanogens.

Methanogens represent a microbial system that can be exploited to produce energy from waste materials.

Large amounts of methane are produced during industrial sewage treatment processes, but the gas is usually wasted rather than trapped for recycling. Extreme halophiles live in natural environments such as the Dead Sea, the Great Salt Lake, or evaporating ponds of seawater where the salt concentration is very high (as high as 5 molar or 25 percent NaCl). The organisms require salt for growth and will not grow at low salt concentrations (Actually, the cells lyse at low NaCl concentrations).

Their cell walls, ribosomes, and enzymes are stabilised by Na^+. *Halobacterium halobium,* the prevalent species in the Great Salt Lake, adapts to the high-salt environment by the development of "purple membrane", formed by patches of light-harvesting pigment in the plasma membrane.

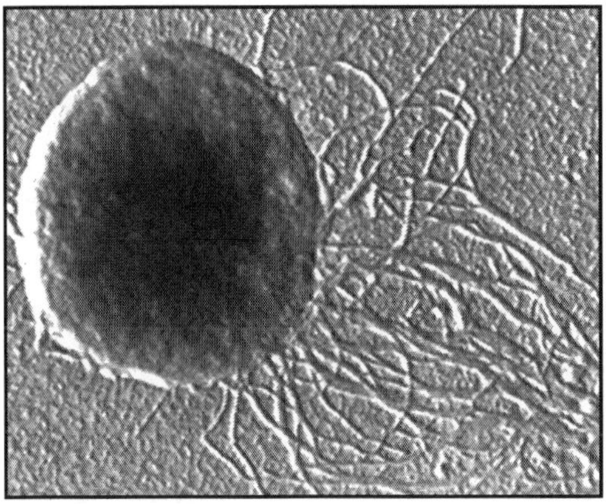

Fig. *Methanococcus Jannaschii* (Holger Jannasch). The Archaean was Originally Isolated from a Sample Taken from a "White Smoker" Chimney at an Oceanic Depth of 2,600 Meters on the East Pacific Rise. It can be Grown in a Mineral Medium Containing only H_2 and CO_2 as Sources of Energy and Carbon for Growth within a Temperature Range of 50 to 86 Degrees. Cells are Irregular cocci that are Motile Due to Two Bundles of Polar Flagella Inserted Near the Same Cellular Pole, making it a Rare Example of a Motile Coccus.

The pigment is a type of rhodopsin called bacteriorhodopsin which reacts with light in a way that forms a proton gradient on the membrane allowing the synthesis of ATP. This is the only example in nature of non- photosynthetic photophosphorylation. The organisms are heterotrophs that normally respire by aerobic means. The high concentration of NaCl in their environment limits the availability of O_2 for respiration so they are able to supplement their ATP-producing capacity by converting light energy into ATP using bacteriorhodopsin.

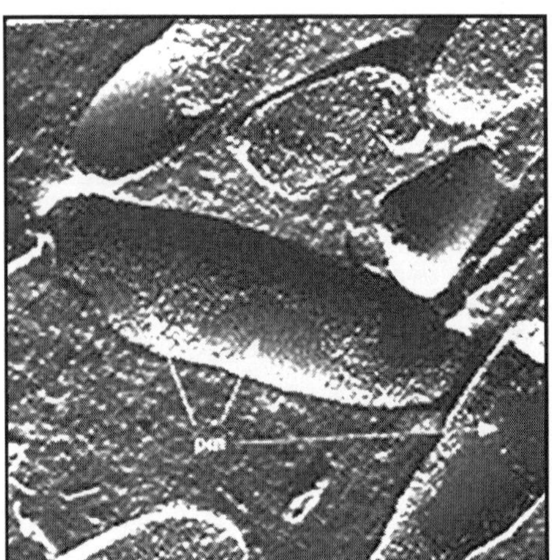

Fig. *Halobacterium Salinarium*is an Extreme Halophile that Grows at 4 to 5 M NaCl and does not Grow Below 3 M NaCl. This Freeze Etched Preparation Shows the Surface Structure of the Cell Membrane and Reveals Smooth Patches of "Purple Membrane" (Bacteriorhodopsin) Imbedded in the Plasma Membrane.

Thermomphiles and extreme thermophiles or "hyperthermophiles" come from several distinct phylogenetic lines of Archaea. These organisms require a very high temperature (80 degrees to 105 degrees) for growth. Their membranes and enzymes are unusually stable at high temperatures. Most of these Archaea require elemental sulphur for growth. Some are anaerobes that use sulphur as an electron acceptor for respiration in place of oxygen. Some are lithotrophs that oxidise sulphur as an energy source. Sulphur-oxidisers grow at low pH (less than pH 2), partly because they acidify their own environment by oxidising S^o (sulphur) to SO_4 (sulphuric acid).

Hyperthermophiles are inhabitants of hot, sulphur-rich environments usually associated with volcanism, such as hot springs, geysers and fumaroles in Yellowstone National Park and elsewhere, and thermal vents ("smokers") and cracks in the ocean floor. *Sulfolobus* was the first hyperthermophilic Archaean discovered by Thomas D. Brock of the University of Wisconsin in 1970. His discovery, along with that of *Thermus aquaticus* (a thermophilic bacterium) in Yellowstone National Park, launched the field of hyperthermophile biology. (*Thermus aquaticus* is the source of the enzyme taq polymerase used in the polymerase chain reaction, PCR., The bacterium has an optimum temperature for growth of 70 degrees.) *Sulfolobus* grows in sulphur-rich, hot acid springs at temperatures as high as 90 degrees and pH values as low as 1. *Thermoplasma, also* discovered by Brock, is a unique thermophile that is the sole representative of a distinct phylogenetic line of Archaea. *Thermoplasma* resembles the bacterial mycoplasmas in that it lacks a cell wall. *Thermoplasma* grows optimally at 55 degrees and pH 2. Interestingly, it has only been found in self-heating coal refuse piles, which are a man-made waste.

The Korarchaeota represent what could be one of the least evolved lineages of modern life that has been detected in nature so far. Barns *et al.* discovered the Korarchaeota in Yellowstone's Obsidian Pool over a decade ago. The group has been subsequently defined only by 16S ribosomal RNA sequences obtained from a variety of marine and terrestrial hydrothermal environments. The 16S-rRNA-based phylogeny of the Korarchaeota suggests that this group forms a very deep, kingdom-level, major lineage within the archaeal domain.

To date, there are no representatives of the Korarchaeota in pure culture, and nothing is known about their properties. However, in recent years, a robust, continuous enrichment culture that harbours these organisms in significant densities has been established. The enrichment community has been phylogenetically characterised and is known to comprise a relatively low diversity of other hyperthermophilic archaea and bacteria. It is thought to be a potential source for DNA that will provide a complete genome sequence for a member of this unique group of procaryotes

Although the Archaea are often inhabitants of extreme environments, there may be corresponding species of Bacteria, and even eucaryotes, in these habitats as well. No bacterium can produce methane, but in many anaerobic environments Bacteria are found in association with methanogens. With regard to acid tolerance, a bacterium, *Thiobacillus,* has been observed growing at pH near 0. A eucaryotic alga, *Cyanidium,* has also been found growing near pH 0. In superheated environments (greater than 100 degrees), Archaea may have an exclusive hold, but Bacteria have been isolated from boiling hot springs in Yellowstone National Park and other parts of the world. No bacterium grows at the highest salt concentration which supports the growth of the halobacteria, but osmophilic yeasts and fungi can grow at correspondingly low water activities where sugar is the solute in high concentration.

Archaeal Cell Envelopes

Archaea possess a broader range of cell envelope structural formats than Bacteria, and their cell walls never contain murein. Some archaea have only a single protein or glycoprotein S-layer as their cell wall (*e.g. Methanococcus jannaschii* and *Sulfolobus acidocaldarius*), whereas others have multiple layers (*e.g. Methanospirillum hungatei*). One archaean has a type of peptidoglycan called pseudomurein (because it lacks N-acetylmuramic acid). Sometimes, there can also be a high proportion of tetraether lipids in membranes, which render the cells more resistant to environmental stresses.

Are Archaea Gram-positive or Gram-negative

Although the Gram reaction depends on both the structural format and the chemical composition of the cell envelope in bacteria, most archaea stain Gram-negative, independent of their basic cell envelope structure or chemical composition. An interesting exception is *Methanobacterium formicicum* that stains Gram-positive, since its cell wall contains pseudomurein, a type of peptidoglycan that lacks muramic acid. In one study, all other archaea stained Gram-negative because their cell walls were so disrupted during staining, that the crystal violet-mordant complex could not be retained by the cells. *Methanococcus jannaschii* was grown at both 50 degrees C and 70 degrees C, which increases the tetraether lipids in its plasma membrane from 20 per cent (50 degrees C) to 45 per cent (70 degrees C) of the total lipids; in both cases the cells stained Gram-negative. One conclusion from these observations is that in a Gram stain preparation of a mixed microbial population containing archaea and bacteria, the archaea are among the Gram-negative cells, and the Gram-positive cells are probably bacteria.

Bibliography

A.P. Narayana Murthy: *Sustainable Livestock Production in India*, Adhyayan Publication, Delhi, 2009.

Amarjeet Singh: *Animal Husbandry*, Rajat Publication, Delhi, 2008.

Aruna T. Kumar: *Handbook of Animal Husbandry*, Indian Council of Agricultural Research, 2008.

Ashok Kumar: *Animal Husbandry*, Discovery Publishing House, Delhi, 2006.

Bibek Ghosh: *Crops Residues and By-Products in Animal Feeding*, Gene Tech Books, Delhi, 2007.

Bibek Ghosh: *Trends and Issues of Livestock Production Systems,* Gene Tech Books, Delhi, 2007.

Birendra Kumar Mishra: *Livestock Production and Rural Development in India*, Biotech Books, Delhi, 2012.

D. Thyagarajan, A. Ashok, G. Raj Manohar and M. Murugan: *Emerging Opportunities in Alternative Poultry Farming Systems*, Satish Serial Publishing House, Delhi, 2013.

Devendra Thakur and D.N. Thakur: *Tribal Life in India, Vol. 3. Tribal Agriculture and Animal Husbandry*, Deep and Deep Publication, Delhi, 2009.

Dinesh Arora: *Biotech's Dictionary of Animal Husbandry*, Biotech Books, Jaipur, 2005.

Gauri Shanker Dokania and Manisha Dokania: *Economics of Dairy Farming : A Micro Analysis*, Satyam Publishing House, Delhi, 2012.

Gyan Deep Singh: *Animal Husbandry*, Anmol Publication, Delhi, 2008.

Harmeet Singh: *Dairy Farming*, APH Publication, Delhi, 2011.

L. Lal: *Animal Husbandry (Objective Fundamentals)*, Agrotech Publication, Delhi, 2010.

Louis M Hurd: *Modern Poultry Farming*, Greenworld, 2003.

M. Singh, D.K. Sharma and U.K. Mishra: *Organic Dairy Farming*, Satish Serial Publication, Delhi, 2011.

Manoj Kumar Rai: *Handbook of Animal Husbandry*, Oxford Book Company, Delhi, 2012.

Manoj Kumar Rai: *Livestock Production and Management*, Oxford Book Company, Delhi,

2012.

Manoj Kumar Rai: *Textbook of Animal Husbandry*, Oxford Book Company, Delhi, 2012.

Mohd. Kalimuddin: *Milk Analysis and Dairy Farming*, Enkay Publication, Delhi, 2012.

N. Das, A.K. Misra, S.B. Maity, K.K. Singh and M.M. Das: *Forage for Sustainable Livestock Production*, Satish Serial Publication, Delhi, 2009.

Nishamani Kar: *Animal Husbandry and Rural Development : Restructuring and Planned Development of the Bovine Economy*, Deep and Deep Publication, Delhi, 2002.

P K Saraswat: *International Encyclopaedia of Dairy Technology and Animal Husbandry (Vols 1 and 2 Set)*, Anmol Publication, Delhi, 2009.

P Mathialagan: *Textbook of Animal Husbandry and Livestock Extension*, International Book Dist, Delhi, 2007.

R Renaville and A Burny: *Biotechnology in Animal Husbandry, Vol. V*, Springer Publication, London, 2008.

R.M. Acharya and Puneet Kumar: *Sustainable Livestock Production*, Satish Serial Publishing House, Delhi, 2013.

Rajendra Reddy and J.P. Abhay Shankar: *Animal Husbandry*, Commonwealth Publication, New Delhi, 2008.

Renaville: *Biotechnology in Animal Husbandry*, Springer Publication, London, 2003.

Shagufta Jamal and H P S Arya: *Participatory Rural Appraisal in Agriculture and Animal Husbandry : A Training Manual*, Concept Publication, New Delhi, 2004.

Shyam Kishore Singh: *Dairy Farming*, Alfa Publications, Delhi, 2012.

Sunil Kumar and B.K. Mishra: *Livestock Production and Management : Recent Trends and Future Prospects*, New India Publishing Agency, Delhi, 2013.

Sunil Kumar, Birendra Kumar Mishra and Manish Kumar: *Advances*

in Livestock Production and Management, Jaya Publishing House, Delhi, 2014.

Suresh K. Rao and Sanjana Rawat: *Economic Importance of Poultry Farming*, Campus Books, Delhi, 2013.

T. Burr Charles and Homer O. Stuart: *Commercial Poultry Farming*, Biotech Books, Delhi, 2011.

Uma Shankar Singh: *Dairy Farming*, Anmol Publication, Delhi, 2008.

V. M. Rao: *Women Self-Help Groups and Dairy Farming*, ABD Publishers, Delhi, 2012.

Vir Singh and Babita Bohra: *Dairy Farming in Mountain Areas*, Daya Publication, Delhi, 2006.

Index